Skype®

FOR

DUMMIES®

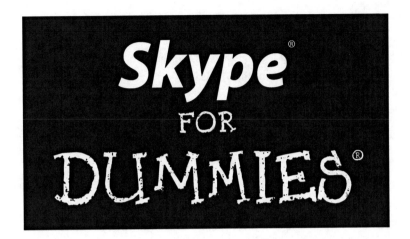

by Loren and Susan Abdulezer and Howard Dammond

Foreword by Niklas Zennström
CEO and co-founder of Skype

BICENTENNIAL
1807
WILEY
2007
BICENTENNIAL

Wiley Publishing, Inc.

Skype® For Dummies®

Published by
Wiley Publishing, Inc.
111 River Street
Hoboken, NJ 07030-5774
www.wiley.com

Copyright © 2007 by Wiley Publishing, Inc., Indianapolis, Indiana

Published by Wiley Publishing, Inc., Indianapolis, Indiana

Published simultaneously in Canada

For general information on our other products and services, please contact our Customer Care Department within the U.S. at 800-762-2974, outside the U.S. at 317-572-3993, or fax 317-572-4002.

For technical support, please visit www.wiley.com/techsupport.

Wiley also publishes its books in a variety of electronic formats. Some content that appears in print may not be available in electronic books.

Library of Congress Control Number: 2006929466

ISBN: 978-0-470-04891-7

10 9 8 7 6 5 4 3 2 1

1B/SX/RS/QW/IN

About the Authors

Loren Abdulezer is CEO and President of Evolving Technologies Corporation, a New York–based technology consulting firm. He is an experienced IT professional serving many Fortune 500 companies. Loren is the author of *Excel Best Practices for Business* and *Escape from Excel Hell* and served as technical editor of *Crystal Xcelsius For Dummies*, all published by Wiley. Loren is *always* exploring new technologies and finding pragmatic and innovative applications. When Skype came along he was quick to recognize its benefits in business and all walks of life. This book is a direct result of wanting to bring those benefits one step closer to a broader audience.

Susan Abdulezer is currently a full time Multimedia Developer in New York City. Susan creates interactive DVDs, documentaries, and Web-delivered media. She has received many honors for technology innovation, winning the prestigious Computerworld/Smithsonian Award in Technology and Academia in both 1996 and 1997. Susan has also written numerous feature articles on education and technology as the contributing editor of Converge Magazine from 1998 to 2002. Susan is active in the Digital Storytelling community, exploring the nature and power of the emerging digital culture. She has also been known to tear herself away from the computer to play classical violin in the Brooklyn Symphony Orchestra.

Howard Dammond is an experienced IT professional and technical instructor, having consulted at several major Fortune 500 companies. Howard has 20-plus years of experience as a technical trainer and developer of innovative learning materials. His perspective on teaching and skills development was first inspired and then intensively developed at Yale University in its unique Master of Arts in Teaching program, where he focused on learning theory, the acquisition and nurture of analytic skills, and interdisciplinary methods of curriculum planning and development.

Dedications

Loren Abdulezer: *To my parents, Ralph and Joyce.*

Susan Abdulezer: *To my parents, George and Cele (better known as CyberPops and Mamou).*

Howard Dammond: *To my wife, Daria; my daughter Rhianna; my son, David; my father, HR; my mom, May.*

Authors' Acknowledgments

In writing this book we feel as though we embarked on an enormous journey. We couldn't have done it without the help and assistance of colleagues, friends, and peers who have gone out of their way to be helpful. We express our heartfelt gratitude and appreciation, and we acknowledge their contribution in the making of this book and its companion Web site (www.skype4dummies.com).

Numerous people have been generous, devoting their time, energy, and expertise. We need to single out two people who were especially instrumental in facilitating the connections to the right people within Skype and throughout the Skype community. Thank you, Kat James and Romain Bertrand. Lester Madden, Dan Houghton, Raul Liive, and Tony Saigh opened many doors for us. We've had numerous and engaging conversations both inside and outside of Skype. In order of first name, we thank the following people:

Aaron Wellman, Adam Gross, Allison Kohn, Anders Hallin, Ash Valeski, Ben Lilienthal, Bernard Percy, Bill Campbell, Bill Good, Brian Phillips, Charles Bender, Christophe Melle, Dani Shefer, David Cohen, David Rivier, Deborah Quinlan, Dick Schiferli, Elspeth Knight, Eric Choi, Eric Partaker, Erica Jostedt, Eyal Gever, Faye Williams, Gershon Goren, Gordon Evans, Graeme Gibson, Grete Napits, Heron Stone, Imogen Bailey, Jaanus Kase, Jen Webb, Jennifer Ruff, Jeremy Hague, Jim Brady, Jin Kim, Joan Gordon, John Martin, John Picard, Karen Gorman, Karen Richardson, Karen Sohl, Kelly Reed, Leslie Schecht, Liz Tierney, Lou Guercia, Martin Dougiamas, Mat Taylor, Melinda Kolk, Natasha Konstantinova, Nicola Riordan, Olivia Selbie, Philip Pool, Philippe Tessier, Phillip Pyo, Rich Conti, Rouzbeh Pasha, Sam Aparicio, Sandy Krochek, Sara Reitz, Scott Miller, Shira Litvak, Stan Kwang, Stella Porto, Stephanie Zari, Tom Gillen, Viktoria Randalainen, and Wendy Dominguez.

We also thank all those friends and family members who helped us put Skype through its paces, and especially George and Cele Pomerantz, who embraced Skype wholeheartedly.

We are grateful for having Susan Christophersen, Leah Cameron, and Colleen Totz Diamond as our editors to give our book shape, clarity, and coherence. Susan Christophersen and Greg Croy did an incredible job of pulling this book together. We also wish to express our gratitude to Jen Webb, Jodi Jensen, Mary Corder, Andy Cummings, and Lisa Coleman.

Publisher's Acknowledgments

We're proud of this book; please send us your comments through our online registration form located at www.dummies.com/register/.

Some of the people who helped bring this book to market include the following:

Acquisitions, Editorial, and Media Development

Project Editor: Susan Christophersen

Acquisitions Editor: Greg Croy

Development Editors: Susan Christophersen, Leah Cameron, Colleen Totz Diamond

Editorial Manager: Jodi Jensen

Media Development Manager: Laura VanWinkle

Editorial Assistant: Amanda Foxworth

Sr. Editorial Assistant: Cherie Case

Cartoons: Rich Tennant (www.the5thwave.com)

Composition Services

Project Coordinator: Kristie Rees

Layout and Graphics: Lavonne Cook, Stephanie D. Jumper, Barbara Moore, Barry Offringa, Laura Pence, Ronald Terry

Proofreaders: Techbooks

Indexer: Jessica Kramer, Techbooks

Anniversary Logo Design: Richard Pacifico

Publishing and Editorial for Technology Dummies

> **Richard Swadley,** Vice President and Executive Group Publisher
>
> **Andy Cummings,** Vice President and Publisher

Mary Bednarek, Executive Acquisitions Director

> **Mary C. Corder,** Editorial Director

Publishing for Consumer Dummies

> **Diane Graves Steele,** Vice President and Publisher
>
> **Joyce Pepple,** Acquisitions Director

Composition Services

> **Gerry Fahey,** Vice President of Production Services
>
> **Debbie Stailey,** Director of Composition Services

Contents at a Glance

Table of Contents

Foreword

∎ ∎

*W*hen we launched Skype back in 2003, our dream was to allow people around the world to talk to each other for free. To make this possible, we created a little piece of software that people could easily download onto their computers and use straight away. Today, talking over the Internet using Skype has become a natural form of communication around the world.

The response to our technology has been amazing. It has exceeded even our wildest dreams. Skype is now used by more than 113 million people all over the world and is available in 27 different languages. And people don't just use Skype to talk to one another. They can do all sorts of things with it — from instant messaging to sending photos to hosting conference calls.

When I found out that Skype was to be included in the popular *For Dummies*, series, I was delighted. This book explains how a good idea can create powerful connections between people and their friends, family, and business colleagues across the world. It also tells you about some of the new gadgets and software you can buy that make Skype really exciting to use. It's incredible to think that only a few years ago, if you wanted to call someone on the other side of the planet, doing so would have cost you a fortune. But now, with Skype, you can call whomever you want for free. And it's fun, too!

We are tickled pink by the way Skype has changed people's lives. And we hope to keep delighting our users just like you every single day. You make Skype what it is. Thank you.

Niklas Zennström

Introduction

"**I**'ll skype you" is quickly replacing, "I'll email you," which ousted "I'll call you." Before any of these, of course, came "I'll send a telegram," which was supplanted by "I'll write you." The need to extend face-to-face communication launched industries around distance messaging, and Skype joins this chorus of change with a "voice" that resonates through the Internet. Skype, however, is not just another way to say hello. Skype combines the power of email, telephony, telegrams, and letters by providing an avenue for text, data, voice, and even video communication. Skype reaches across oceans but has tools to make the experience intimate. You can exchange files as easily as handing someone a piece of paper. You can see eye-to-eye through video conferencing. You can convene a group of friends with ease. Millions of people are discovering all this versatility with Skype. *Skype For Dummies* makes it easy for you to join them.

About This Book

Skype is simple to download and start using, so why write an entire book about it? There's much more to Skype than simply calling someone from one PC to another, and *Skype For Dummies* opens your eyes to the many ways in which you can put Skype to use, with exciting possibilities for both your personal and professional development.

As is true for the world of technology in general, Skype technology is a rapidly moving and dynamically evolving target. So some of the software or gadgets we describe here may have already changed between our writing and your seeing this book in print. But we have tried to capture the spirit of where the technology is headed, and *Skype For Dummies* is loaded with the information and ideas you need to get skyping right away.

This book shows you how to

- Get started if you've never used Skype and become an expert in no time
- Find Skype communities to join
- Pick out just the right gadgets for optimum skyping
- Discover Skype-based solutions for your business
- Explore, understand, and adopt a new set of ideas about communicating

How to Use This Book

Jump right in anywhere! We've designed this book so that you don't have to read it from front to back; it's a user-friendly reference tool that you can even start reading at the back if you're so inclined — you won't spoil the ending.

Foolish Assumptions

In writing this book, we've made the following assumptions about you:

- ✔ You wouldn't mind making free or low-cost calls.
- ✔ You can use a telephone, computer, or mobile device.
- ✔ You are curious about technology and wouldn't mind trying a few new things.
- ✔ You don't want to be left out of the communications revolution.

How This Book Is Organized

We've organized this book so that you can easily find whatever you need or want to know — whether you're new to Skype or are fast becoming a veteran skyper.

Part 1: Getting Started with Skype

The first chapter in this part provides a snapshot of what Skype is all about. If you're ready to get skyping now, you can plunge in to Chapter 2 for everything you need to know about downloading Skype and setting it up to make your first call. In Chapter 3, we take you sightseeing through the Skype interface, detailing its menus and toolbar. As you continue to explore the world through Skype, use this chapter as a quick reference for whatever task you need Skype to do for you.

Part II: As You Like It: Skype Your Way

We've found that the more you fine-tune your Skype settings to suit your daily needs, the better Skype will serve you. To that end, Chapter 4 is full of information on how to make Skype behave concerning notifications and alerts, sound effects, protecting your privacy, and much more. Chapter 5 describes ways for you to broadcast your personality to the world, both for fun and as a business communications strategy. In Chapter 6, you can find out all about online chatting through Skype. Chapter 7 launches you into the exciting world of skyping with video and shows you how to choose the best webcam for your purposes. Finally in this part, Chapter 8 gives you the scoop on connecting Skype with ordinary phones using services called SkypeIn and SkypeOut.

Part III: Calling All Seasoned Skypers

Even if you're not a seasoned skyper yet, browsing this part of the book may make you eager to become one! Chapter 9 covers a host of ways to send and receive messages, perhaps in ways you've never imagined. And Skype makes conferencing easy, whether with one or many others and whether for business or social activities, as Chapter 10 reveals. In Chapter 11, we unveil a gallery of gadgets and add-ons that you can use to greatly enhance your skyping.

Part IV: The Professional Skyper

Skype is a heavy-duty communications engine for commerce, and Chapter 12 helps you consider whether it's time to join with the millions of business users who are already benefiting from Skype. But "professional" skyping doesn't just imply business-related topics, and Chapter 13 gives you a taste of the array of online communities springing up all over the world through Skype. Last in this part, Chapter 14 tells you how to both join and create your own Skypecasts, which are free, large-scale Web conferences.

Part V: The Part of Tens

We had fun brainstorming our lists of "ten things" in each of these chapters, but they are also very practical. You might want to start with this part first, just to get a sense what Skype is and how it's helping to change the world.

Conventions Used in This Book

We use some conventions throughout this book that merit a little explanation. When you see a phrase such as "choose File➪Edit My Profile," it means to click through a given sequence of menu commands. In this example, those commands are File followed by Edit My Profile.

Whenever we tell you to click something (most likely it's a button or an icon), you use the left mouse button and click just once. On those rare occasions when clicking twice is required to get the job done, we tell you to double-click.

To *select* an item, you either highlight it or click in a check box or radio button, depending on the item. Text that we tell you to enter (that is, type) into the program, such as in a text box, appears in **boldface** type. Web site addresses and on-screen messages show up in `monofont` type. To signify hyperlinks, otherwise known as just plain links, we underline the text of the link. On your computer, clicking such a link transports you to another location altogether, such as a Web page.

Finally, to avoid confusion, we use title-style capitalization for option names and links when they appear in regular text, even when the program doesn't.

Icons Used in This Book

To flag special points we want to draw your attention to, we use the following icons:

This icon shows up next to a special tidbit of information or a trick to make something easier.

This icon indicates information that we want to make sure you notice.

Text next to this icon tells you something that you need to, well, remember.

This icon signifies geeky stuff that may interest some, but certainly not all, readers.

 When you see this icon, take heed; it signifies something you need to be aware of before you act.

 This one doesn't show up very often, but it means be *very* careful — the stove is hot and you can get burned.

Where to Go from Here

Without further ado, we urge you to start skyping. As you experience this extraordinary way to connect to the world, pay us a visit from time to time at the Web site for this book. We'll be posting updates, tips, tricks, new gear, stories, and more. Find us at

```
http://www.skype4dummies.com
```

Part I
Getting Started with Skype

In this part . . .

This is the place to begin if you're not exactly sure what Skype is, where to find it, or how to get started using it. In this part, you get a brief introduction to the world of Skype, find out how to download it, sign up for a Skype Name, locate fellow skypers, and get talking! You also get the lowdown on navigating through the Skype menu with all its options and tools.

Chapter 1

What's All the Hoopla about Skype?

Do you know what happens when you speak to someone over a phone line? Your conversation is converted into an electronic signal and sent over a copper wire or some wireless network to someone else's phone. The phone companies have set up vast networks to seamlessly connect you to just about anyone on the planet and these days, the networks are all digital even if your phone is not. In addition to managing the call, phone companies track where you are dialing to and how long the call persists so that they can send you a bill at the end of the month.

Wait a second — if the zeros and ones pushed through the phone lines are the same as the zeros and ones found on computers like yours, networks, and the World Wide Web, why can't you push those zeros and ones through the Internet? Well, you can. That's what Skype is all about.

In this chapter, you find out what makes Skype different from regular telephones, along with surprising ways to use Skype and a quick overview of ways to make Skype fun and productive.

Seeing What Skype Can Do For You

Skype can dramatically alter how you exchange information, how you meet new people, and how you interact with friends, family, and colleagues.

Although you can make calls on Skype, there is oh so much more to it. For starters, here are some things you get or can do with Skype:

- Call to or receive a call from a regular telephone, a cell phone, or a computer on the Internet.
- Send or receive files over the Internet to and from fellow skypers.
- Search your Outlook contacts and call them within Skype.
- Search the Skype database of all Skype users on the planet.
- Hold a conference call with a group of people. Besides participating in audio conferencing, you can "simul-chat" with your conference participants — exchanging text, live Web links, and files.
- Make live video calls.
- Initiate a group chat.
- Hold a Skypecast for as many as 100 people at a time.
- Transmit secure and encrypted voice conversations, text, file transfers, and video.

Skype (the basic stuff) is free

To use Skype, you need only three things:

- **A computer with access to the Internet:** Your Internet connection should be faster than dial-up. Just as Web access with dial-up does not work very well, the same is true of Skype with a dial-up connection. You're best off using a high-speed broadband connection DSL or cable
- **A free software program called Skype:** You can get this program from www.skype.com (see Chapter 2).
- **A microphone and headset:** Plenty of audio device options are available (see Chapter 11), ranging from inexpensive ($20 or so) to a little more pricey.

That's it. To be able to make your first call, you just download the Skype software from the Internet, create a Skype Name for yourself, test your audio connection through Skype, and you're good to go. You can talk to fellow Skype users around the world without any time limits and without having to pay anyone.

If you want more than just the basic service, however, you will have to pay. Skype is free when you talk to another Skype user on a PC. But what if you want to call, say, your grandmother, who doesn't have a computer? For a small fee (as low as two cents a minute or even free), you can "SkypeOut" from your computer to a conventional phone. The cost depends on whether she's in the same country as you are and how long you speak. Similarly, someone who doesn't have Skype can call you using the "SkypeIn" feature. (See Chapter 8 for more about using SkypeIn and SkypeOut.)

You can use equipment and services you already have

Why would you want to use the Internet to manage phone conversations? First, you already have it (and pay for it). Internet use is widespread; it seems that nearly everyone has it. The technology keeps improving every day. If you already have access to the Internet, you can handle much of your long-distance calling over the Internet for very low cost or for free, and often with better sound quality than you get from cell phones and regular phones. Also, you may be able to do things with an Internet-based phone system that you can't do with a conventional phone system. You can run a Skypecast with a hundred people. You can send and receive files that would choke your email system. You can send and receive live video from anywhere in the world. Better yet, all these transmissions over Skype occur on secure and encrypted lines of communication.

It's great for personal communications

The Skype community is international. People from all corners of the globe show up in the searches, so it helps to provide identifying information in your Skype profile, such as your language, country, and perhaps city.

Click the Profile icon to get public information about any Skype user selected in your search results (see Figure 1-1).

Figure 1-1:
The Skype
Profile page.

More important, you can add a person to your Contacts list (see Figure 1-2). Adding someone to your Contacts list entails seeking that person's permission and receiving his or her contact information. See Chapter 3 for more details about contacts, including how to find and add them.

Figure 1-2:
Adding a
Skype user
to your
Contacts
list.

You'll quickly find that people enjoy working with Skype and are usually eager to exchange information on how to best use Skype.

After you start making the rounds, you'll want to try the chat or instant messaging capabilities. You can find plenty of information about chats in Chapter 6.

And great for professional services, too

Chats are fun, but Skype can be good for your career, too. Think about how Skype can serve you. Say, for example, that you're sitting in an airport waiting for your flight. The Director of Sales has just attended a briefing with a major customer and found out that the customer needs double the amount of merchandise that was negotiated six months ago. Now the two of you need to validate that you can fulfill the revised production schedule, and doing so involves the Product Development Manager. Why not set up an impromptu conference call that includes all three of you? You're in an airport in Chicago, which has a public Wi-Fi network. The Sales Director is in the Dallas office, and the Product Development Manager is at home in San Diego. But that's okay — it doesn't matter where you are.

One of the marvels of this technology is that any of you can instantly convene a conference call, and you can all be in different cities around the globe. It is great to have the convenience of a conference call. This one sounds as though it's pretty serious. Think you might need to exchange documents, such as spreadsheets or PowerPoint slides? Well, you can do just that. To find how to transfer a file, go to Chapter 6.

Hold on — what about the fact that you're sitting in an airport? An airport can be a favorite spot for industrial espionage. Skype, however, creates a roadblock for your corporate competitor: All the file transfers, all the chats, and all the conversations that take place are secure and encrypted!

Going Beyond the Basics with (Not Necessarily Free) Services

Skype does many things right out of the virtual box. But you can add a full complement of features and services provided by Skype and third-party companies.

Skyping in from a phone

How can people call you from a regular telephone if Skype runs on a computer? Well, for a small fee you can get a SkypeIn phone number that anyone with a regular phone can call. There are no surcharges of any kind. The person calling you calls your SkypeIn number. As far as he or she is concerned, the call is being made to a regular phone number. The charges incurred by the call amount to whatever they would be to make a call to a phone in your area code. When you sign in to Skype, SkypeIn calls are automatically routed to you. You can be anywhere on the planet. If you are connected to Skype, the calls get to you just the same.

Skyping out to a phone

Not only can people call into Skype using an ordinary telephone, but you can call out to anyone on a landline or mobile phone using SkypeOut. Making a call from a computer to a telephone is as easy as entering the phone number and pressing Enter. The charge for the call appears on your screen (if there is a charge; some calls are free). How do you pay for this? Buy a block of minutes with SkypeOut credits. When your SkypeOut minutes are used up, you can purchase more. Chapter 8 gives you all the information you need to get started with SkypeIn and SkypeOut.

Skype terms

These terms may not have made it into the Oxford English Dictionary yet, but who knows, maybe you can say you saw them here first!

to skype (verb, infinitive form): To communicate over the Internet using voice, video, text, and file transfers using a special program called Skype.

skyping (present participle): The act of calling over Skype.

skyper (noun): A person who skypes.

Skype (proper noun): The software and the company that makes it.

Skype Name: Your personal Skype ID.

SkypeIn: A service provided by Skype through which a person using a regular phone can call and connect to a skyper.

SkypeOut: A service that allows a skyper to call a regular telephone using the Skype software.

Skype Me: A mode of alerting the world that you are available for contact and making new friends.

Getting voicemail

What's phone service without voicemail? You can add voicemail to Skype with Skype Voicemail (a plain-vanilla service), or with Pamela or Skylook (both are packed with features). Skype Voicemail is bundled with SkypeIn. If you want Skype Voicemail without SkypeIn, you can purchase it with Skype credit. Pamela and Skylook have "lite" versions for free, but the good stuff costs a few dollars (really, just a few).

See Chapter 9 for more about Skype Voicemail, Pamela, and Skylook.

Getting a Load of Add-Ons and Accessories

Having more features and capabilities to enhance Skype can be highly desirable, and Skype makes it easy to integrate third-party applications and plug-ins to work with its software. Skype permits and in fact encourages third-party companies to bring enhancements to Skype. (Such enhancements are handled through something called Skype Application Programming Interface [Skype API], but that's more than you need to know.)

Going wireless

Skype delivers a new kind of freedom — the ability to make and receive calls wirelessly and without even needing a computer. A new type of phone, called a Wi-Fi phone, has Skype already built in and can tap into wireless or Wi-Fi networks. When you're connected to a network, whether from your office, home, or public Internet hot spot, you can use Skype. With these devices, you sign on to Skype as you would from your computer and you're good to go.

Software and gadgets, too

When you connect through Skype, you need some way to speak and hear your conversations on Skype. Most computers have an audio jack for a microphone and a speaker out line. If you don't already have a microphone and headset, you can purchase them inexpensively. Bluetooth (wireless) headsets, USB headsets, USB speakerphones, mini-plug headsets, and built-in microphones are available in all colors and sizes, some with noise cancellation, volume control, and mute buttons. You have many choices.

The capabilities of Skype are not limited to traditional "telephone" calls. Skype supports videoconferencing, for which you'll need a good webcam. You can record interviews and save them as media files with Skype, so you'll need recording and playback software. If you like to skype but don't like to be stuck at a desk, you can skype from digital handsets known as Skype Phones that work within 50 feet of your computer. You can even store your whole Skype operation on a USB drive, travel the world, and simply plug and play wherever you land. Chapter 11 covers a galaxy of Skype gadgets and software, so get ready to go where no one has gone before!

Making Skype Play Well with Others

Unlike telephones, Skype is a social butterfly capable of connecting a handful of people for a conference call, up to a hundred people for a chat, or a hundred people for a Skypecast. But as with all social butterflies, you have to set down some rules. These rules and guidelines are spelled out in the various chapters on Skype conferencing (Chapter 10), chats (Chapter 6), and Skypecasting (Chapter 14).

When enterprise security matters

Skype can pass text, data, voice, and video through firewalls and routers without requiring special setup. For this reason, Skype can bypass normal IT security controls. However, Skype can be deployed and configured to prohibit file transfers over Skype and using the Skype API. Chapter 12 provides some suggestions on how to deploy Skype in this locked-down mode.

When the Web is your playground

Skype mingles easily with social, business, and e-commerce networks, such as Bebo and eBay (see Chapter 13). Skype social communites of all stripes — for dating and friendship, education, special interests — are connecting people from all over the world.

When you sign on to these Web-based communities, your Skype Name is posted along with your email and any other contact information you provide. On these online communities, a little icon next to your Skype Name changes when you are online, offline, or just don't want to be interrupted. But the best part is that although everyone has your Skype Name, you can require people to ask permission to speak with you (see Chapter 4). A little different from annoying sales calls at dinner, isn't it?

Chapter 2

Hooking Up with Skype

Setting up Skype and getting it working for you is not difficult, and you can do it using the quick steps presented in Chapter 1. Knowing about the process in greater detail, however, will help you get comfortable with Skype more quickly. That's what this chapter is for. Also in this chapter, you find out how to install Skype and create a new account; how to fill in your profile to let the world know who you are; how to find others on Skype; and, best of all, what to do to make a Skype call and start speaking to the rest of the world through your computer.

As mentioned in Chapter 1, the Web pages on www.skype.com are constantly being enhanced. So, they may change from time to time. Don't be surprised if the Web pages you're looking at don't exactly match the screen shots in this book.

Downloading and Installing Skype

Ready to get started? First, pick up the latest version of Skype:

1. **With your Internet browser open, enter www.skype.com in the address line to open the Home page of the Skype Web site.**

2. **Click the Download button on the Skype home page to open the Download page.**

3. **Click the link for the computer platform that you use to download the software and then click the word Download immediately under the name of the platform you just clicked (see Figure 2-1).**

   Skype works on Windows, Macintosh, Linux, and Pocket PC. In most of the examples in this book, we use Windows, but we also discuss, where appropriate, differences with other platforms.

4. **Click the Download button (see Figure 2-1) to download Skype to your hard drive.**

   The appearance of the download page and download button varies by computer platform. On the Mac and Skype for Pocket PC, the button says Download Now. You have to pick the particular variant of Linux you want to download and then save the software to your hard drive.

   After you click to download the program, you may be prompted to save or open the program SkypeSetup.exe. *Always* choose to save the file to your hard drive. Do not open or run the program during the download process.

   Skype often posts two types of downloads, the general public version and the next new beta version. If you are just starting out, it's best to download the general public version because it's more stable. In time, the beta version will become the public version, and you can upgrade.

Figure 2-1:
Click the
Download
button to
download
Skype.

5. Select Save to Disk.

Choose an easy-to-remember location, such as the Windows Desktop, and click OK. Remember this location so that you can double-click the file after it is saved.

When your download is complete and the program is saved to disk, follow these steps to install the program:

1. Double-click the SkypeSetup program that you just saved.

If you are on Windows, you will see a little blue-and-white cube that says SkypeSetup.exe. If you are on a Mac, you will see a standard installer icon that looks like a disk drive but is colored blue and white with a big Skype *S* on top.

2. When the installer program launches, it asks you to choose a language; select it from a drop-down list (see Figure 2-2).

The installer program then asks you to read and accept the End User License Agreement. You need to accept this to continue the installation.

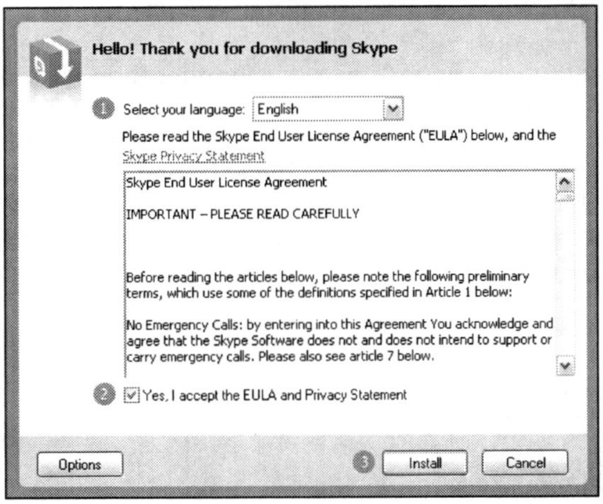

Figure 2-2:
Choose among many languages for the one you want to tell Skype to use.

3. Click the Options button (refer to Figure 2-2).

The Options page that results (see Figure 2-3) asks you to select where Skype should be installed. In the field provided, enter the path to the selected location. You can also click Browse to simply select a location to install Skype.

You also have the choice to launch Skype as soon as the installation is complete. This box is already checked. If you *don't* want Skype to start up right away, deselect the box by clicking it.

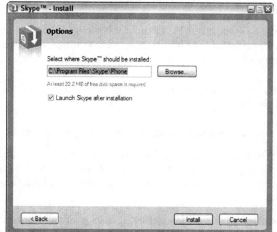

Figure 2-3:
Customizing
your Skype
installation.

4. Click Install.

Skype is installed and automatically launches (unless you deselected the Launch Skype check box on the Options screen).

 Just in case you want to use Skype in a language other than English, you have a considerable range of options with a growing list of languages, including the following: Arabic, Bulgarian, Chinese Simplified, Chinese Traditional, Czech, Danish, German, Estonian, French, Greek, Hebrew, Hungarian, Italian, Japanese, Korean, Norwegian, Dutch, Polish, Portuguese, Brazilian Portuguese, Romanian, Russian, Spanish, Finnish, Swedish, and Turkish. (Try saying that all in one breath!)

Choosing Your Skype Name and Password

When you launch Skype for the first time, you see the Sign In to Skype window, shown in Figure 2-4.

To use Skype, you must set up a Skype Name and password. A *Skype Name* is what other people on Skype see when they contact you. You use your Skype Name to log on to Skype in combination with a password. Because you choose your Skype Name, pick one that's unique to you and easy to remember.

Figure 2-4:
Launching
Skype for
the first
time.

To establish your Skype Name and password, follow these steps:

1. **Click the <u>Don't have a Skype Name?</u> link (refer to Figure 2-4).**

   The Create a New Account page opens.

2. **In the Enter Skype Name box, enter the name that you've decided to use.**

   One of the first questions that may pop into your mind is, "How will all my friends and family find me on Skype?" If you already have a screen name that you use for your email, try using the same name or something very similar on Skype. However, be aware that Skype Names must start with a letter and not include spaces. Skype Names also must be between 6 and 32 characters long, including letters and numbers. All Skype Names are converted to lowercase letters, even if you use uppercase letters when you create your account.

   What if you have a great idea for your Skype Name but somebody else beat you to it and registered the name on Skype? For that possibility, Skype offers you alternative names to pick from — but also allows you to set one of your own choosing (see Figure 2-5).

3. **Choose a password and enter it in the Password box.**

   A Skype password must be at least four characters long. All Skype passwords are case sensitive.

   You are asked to enter your password again in a second Password box to make sure that you entered it accurately.

Figure 2-5:
Choosing an
alternative
when your
preferred
name is
taken.

Pick a password that is easy for you to remember but difficult, of course, for others to guess. A good technique is to pick two short but unrelated words and separate them by some nonalphabetical characters — for example, latte67chair.

4. **Select the check box to accept the Skype End User License Agreement and the Skype Privacy Statement.**

   Although you agree to the License Agreement when you download Skype, you must check it again when you create an account before you can successfully log on to Skype.

5. **Click Next to continue to another Create Your Account Screen; there, enter your email address.**

   Entering your email address is optional but recommended. Be sure to fill this in. Email is the only way to retrieve a lost Skype password. Also, if you don't have email, you'll have a difficult time making use of features such as SkypeOut and SkypeIn (see Chapter 8).

6. **Enter your country, region, and city.**

   This information is optional, but it's useful when other skypers search for you.

After you complete the Skype Name information, a page that looks much, if not exactly, the same as that shown in Figure 2-6 appears. Click the Sign In button to launch Skype. Skype is now live and ready to use, and you are greeted with a helpful Getting Started guide. We encourage you to take the time to view this simple tutorial, although you can find out how to perform all these tasks in this book as well. The guide walks you through the general steps for:

✔ Testing your sound settings

✔ Adding a Skype Contact

✔ Making a Skype call

If you want to refer to the Getting Started guide again, you can open it at any time through the Skype menu by choosing Help⇨Getting Started.

Figure 2-6:
Give
yourself a
Skype
Name and
sign in.

What's your Skype handle?

Choosing a Skype Name that represents your personality, a character trait, or physical trait recalls the "handle" craze in the 1970s, when Citizen Band, or CB, radio became popular. CB radio owners picked unique names, known as "handles," to identify themselves to other CB radio operators.

CB radio was a free radio frequency set up by the Federal Communication Commission in 1947. CB radio is limited to a ten-mile radius — big enough for plenty of conversation on a highway filled with truckers. As with Skype, CB radio operators loved the fact that the calls were free.

We wouldn't be surprised if many of those passionate radio hams decide to use their old handles with this new version of the "people's phone system," Skype.

Your Skype "handle" is an opportunity to invent your own branding. What you decide to call yourself should also be influenced by whether you plan to use your Skype Name for business, personal, or both purposes. In considering your Skype Name and how you fill out the whole Skype profile, strive for balance between formality and self-expression.

Filling Out Your Skype Profile: Making It Profile You

Every new Skype member has an opportunity to enter information into a personal profile screen. You don't *have* to fill in any information; profiles are completely optional. The Skype Profile window clearly states that whatever details are included are "details that only my contacts will see" (see Figure 2-7). So, don't feel obligated to fill in the blanks!

There are some good reasons, however, to populate those empty profile fields. The benefits of adding information are to

✔ Help others find you in a Skype search if they don't have your Skype Name

✔ Give clients easy access to telephone contact numbers that you make public

✔ Show new friends your photo

✔ Post an image of your business logo

✔ Provide contacts with a link to your Web site

✔ Show your local time so that others know when to contact you

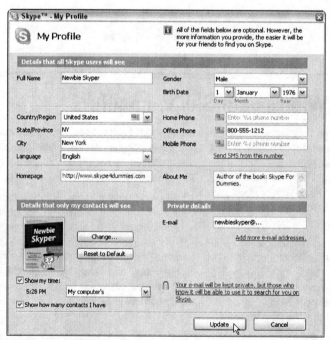

Figure 2-7:
How you fill out your Skype Profile is how the world will see you.

Your Skype Name is not part of your profile. A Skype Name is not optional. You need it to log on to Skype and to enable other skypers to contact you. Your profile details give more information about who you are and what you do. You can let the world know whether you are a businessperson, an artist, an author, a fiddler, or just a seasoned skyper looking for a few friends to hang out with online. As more and more people use Skype for business, special interest groups, educational venues, and social gatherings, their location, language, and other personal information may form the basis for a common interest or mutual benefit. Personalizing Skype is more than self-expression; in some cases, it is a courtesy to others reaching out to you.

To open your Skype Profile screen, choose File⇨Edit My Profile from the Skype menu. You have the option to fill in the following fields:

- **Full Name:** The first bit of information you can publish to the world in your personal profile window is your full name. Many of us shy away from offering our real names to the virtual world; others find it liberating to invent an identity; and some prefer to be completely anonymous. However, there are advantages to adding your real name to your profile:

  - Adding your full name makes it easy for others to find you on Skype.

  - If you use Skype for business, your full name is a vital piece of information for clients to access.

  - In a conference call or chat, any member of the group can look up your profile to find your name. If many people are conferencing or chatting for the first time, forgetting a name is easy, and it's handy to have it available.

- **Gender:** Potential dates certainly want to know this fact. If your first name is slightly ambiguous (think of all those male Leslies, and Lorens, one of the authors of this book), filling out your gender might just turn out to be the happiest bit of information you share with the world.

  Gender is another option that can help you (or others) narrow a search in the Skype directory. If gender is not included in a person's profile, it is not searchable.

- **Birth Date:** Most people are reluctant to reveal this detail. There are, however, some reasons to include your age, such as if

  - You are open to meeting people your own age.

  - You are an advocate for a specific age group, such as teens, the elderly, or baby boomers.

✔ **Country/Region, State/Province, or City:** Skype is international. It's important for callers to have some idea of where they are calling. Knowing where someone lives, and letting others know where you live, can help you both prepare for cultural differences in communication. Whether used for business or social reasons, this is an essential piece of information for smooth customer relationships. In addition,

- All locations are searchable in the Skype directory.

- Location provides information on local time zones. If this information is not included elsewhere in the profile, at least you'll know when to call (and when not to call).

✔ **Language:** Choose your language from a drop-down list.

- Indicating your native tongue lets callers know whether they share your language. If they don't, they may choose to engage a translator in advance.

- Language is a searchable item in the Skype directory.

✔ **Home Phone/Office Phone/Mobile Phone:** Skype allows you to list three phone numbers, one for home, office, and cell. Some good reasons to add these numbers are the following:

- Your Skype profile is available worldwide, so your phone numbers have a global reach. If your phone number is listed in your local phone directory or your business is listed in the Yellow Pages, you would do well to place the equivalent listing in your Skype profile.

- You can update your telephone information on Skype instantly. This feature enables you to provide the most current contact information, even if you are in the middle of a major relocation. Updating local directories, or even Web listings, takes much longer.

- If someone cannot contact you on Skype, your profile gives that person options in an emergency. People can SkypeOut to you from within the Skype program, or just use a cell or landline. In either case, they have instant access to your phone numbers.

✔ **Homepage:** It costs nothing to create a Skype account and list a Web page that you want all the world to see. The Homepage box has room for only one Web address, but it can be an effective contact tool for you because:

- Web site links are active. A contact can navigate to your Web site directly from your profile.

- Posting a Web site in your profile is free advertising for your business, product, service, or just yourself.

✔ **About Me:** This is your chance to tell your story in 200 characters or fewer (note that it's 200 *characters*, not words). Plenty can be said in a few words. You can

- Describe your business.

- Put in a short bio.

- List your hobbies, collections, or favorite activities.

✔ **Adding a photo or artwork:** You can add a 96-x-96–pixel graphic to your profile. The format for your graphic must be JPEG or BMP. You can add a picture in two ways:

- Click the picture box in the lower-left corner of your profile window. A My Picture screen opens. Click Browse to search for a picture you have prepared in your computer.

- Click the Change button to the right of the picture box. Doing so also brings you to the My Picture screen. Click Browse and search for a picture.

Your picture is visible only to your Skype contacts, that is, only people you have in your approved contact list. People who find your Skype profile in a search will not be able to see your picture until you give them permission (see Chapter 4).

You may have hundreds of approved contacts, but you may post only one picture. Pictures make your Skype profile more personal. They can identify you with a business, club, or charity. They can add humor. Some pictures on profile pages include:

- A business logo

- A recreation picture, such as skiing, surfing, or golfing

- A head shot

- A cartoon

- A favorite landscape

See Chapter 5 for more about including graphics with your profile.

✔ **Show My Time:** Skype is available 24 hours a day in all time zones. Letting your contacts know your local time or time zone (only your contacts will see your time zone) is important. Your profile page enables you to do indicate your time zone in two ways:

- Click in the Show My Time box. This action displays the actual time of day as shown on your watch or wall clock.

- Select a GMT (Greenwich Mean Time) zone from the drop-down list provided (see Chapter 10 for more details).

✔ **Show How Many Contacts I Have:** Unless you tell it to do otherwise, Skype automatically displays how many Skype contacts you have. Although it is nice to share how many contacts you have, you may not want to convey this information. Simply deselect the Show How Many Contacts I Have check box.

✔ **Email:** You can provide up to three email addresses. These are completely private; no one, not even your contacts, can see these on your profile. Even if no one can see them, there are two reasons to include email addresses, as follows:

• If you forget your Skype password, you will need your email address to retrieve it.

• People who have your email address can use it to search for your Skype Name, even if the email address does not appear in any Skype directory.

One great activity is to look up your own last name in the Skype search window and see whether you find distant family members in other countries. Some delightful surprises come out of this activity. In fact, we (Loren and Susan) recently found a cousin from England we'd never met and we spoke for the first time over Skype.

Welcome to the skyper clan

In our example screen shots used throughout this book, we show examples with various Skype Names. In the early chapters, NewbieSkyper appears often in our screen shots and examples. Later chapters contain numerous screen shots of SeasonedSkyper. You'll also see TheProfessionalSkyper and CuriousSkyper.

 I'm NewbieSkyper. Like yourself, I'm new to Skype. I've heard about this cutting-edge stuff called Skype and thought that using it would be hard, but it's much easier than I thought.

 I'm SeasonedSkyper. Skype has become an integral part of my life. I have friends and relatives overseas whom I call regularly. With Bluetooth (see Chapter 11), I am no longer tethered or wired to my computer. I just started using a webcam with Skype. I am hooked!

 I'm TheProfessionalSkyper. I train my clients using Web conferencing in conjunction with Skype. This makes it easy to use applications, discuss business strategies, transfer files, and log communications with my partners around the globe.

 I'm CuriousSkyper. I like to call people I don't know and find out whatever I can.

You have a number of available options for setting the time that will be displayed. Generally, it is easiest and best to set the time to match your local computer. If you are traveling for the week and will be in an entirely different time zone, you can manually set the displayed time to match a specified time zone. For example, if you reside on the east coast of the United States, the local time zone is –5:00 GMT. If you are traveling to San Francisco, which is on the west coast of the United States, you can set your time zone to –8:00 GMT, because it is three hours earlier in San Francisco. When people view your name on the Skype Contacts list, they see the time associated with where you are traveling.

Testing Your Connection

Before you talk to someone, you should test your audio equipment to make sure that you have it properly connected and the volume levels are set appropriately.

To test your connection, follow these steps:

1. **Double-click your Skype icon on the desktop or choose Start⇨ All Programs⇨Skype to launch Skype.**

   The main Skype window, which simply shows "Skype" followed by your Skype Name across the top, appears and looks similar to the one shown in Figure 2-8. If for some reason the main Skype window is obscured, repeatedly press Alt+Tab on the PC or Option+Tab on the Mac to cycle through all open applications until you navigate to the Skype application.

Figure 2-8: The main Skype window appears when you launch Skype.

2. **Enter** echo123 **in the text entry box that contains the text "Type Skype Name or number with country code."**

   Normally, this box is where you would enter the Skype Name or number of the person you want to call, but entering **echo123** connects you to the Skype Call Testing Service.

3. **Click the large, round green icon that appears just below the text entry.**

   The green call button represents a handset and is the one you click to place a call. (The red hang up button next to it is for hanging up or closing calls.) Clicking the green button connects you to the Skype Call Testing Service (see Figure 2-9). When you're connected, you hear a personable voice calmly guiding you through an audio test.

4. **Leave a short message when prompted, up to ten seconds long, and listen for your message to be repeated.**

   If you don't hear your message, something's wrong!

The Skype Call Testing service allows you to hear how you sound when you call others. Perhaps your volume is too soft or too loud, or nonexistent. This is a good time to set things right before you attempt your first Skype call. Here are some quick tips and suggestions for troubleshooting your sound problems. The end of this section provides a link to a more comprehensive troubleshooting guide on the Skype Web site.

If you have sound problems when making a Skype call, try the following:

✔ **Make sure that your headphone and microphone jacks are plugged into the right audio in and audio out sockets on your computer.** Try reversing the order.

✔ **Compare your Windows Sound settings with your Skype Sound settings.** To locate your Skype Sound settings, choose Tools⇨Options⇨ Sound Devices. To locate your Windows Sound settings, choose Start⇨ Control Panel⇨Sounds and Audio Devices. (This path may differ on some computers with different operating systems and configuration setups, but your aim is to locate the Sounds and Audio Devices control panel.) Make sure that the sound devices match.

✔ **If your voice sounds very quiet to others over Skype, first try moving your microphone a little closer to you.** If moving your microphone doesn't work, try this: Choose Tools⇨Options⇨Sound Devices and deselect the check box next to Let Skype Adjust My Sound Device settings. Skype no longer automatically sets the volume, thereby allowing other programs, including your operating system, to set the volume for sound input and output.

✔ **If no one can hear you, you may have the wrong device selected in your Skype sound settings.** If more than one device is listed, make sure that the device selected is the microphone you are using. For example, if you're using a microphone that connects through a USB port, be sure that you've selected a Sound setting that tells your computer to find the mic in that port.

✔ **If you can't hear anything, you may have the wrong audio out device selected in your Skype sound settings.** If more than one device is listed, make sure that the device selected matches the headset or speakers you are using.

You can find advanced help in setting up your sound devices at www.skype.com/help/guides/soundsetup.html.

Figure 2-9:
Use the
Skype Call
Testing
Service to
pave the
way for
smooth
calling.

You can hear the same call-testing service in Chinese if you skype echo-chinese (include the hyphen) instead of echo123. Another one is soundtestjapanese for Japanese.

Eliminating feedback

You may encounter some feedback problems when using Skype, but these have simple solutions, as the next sections explain.

Feedback you can hear

Even if you are prepared with the appropriate headgear for Skype calls — meaning that your headset/microphone unit is plugged into your computer — the person you are calling may not be as prepared. If you skype someone who is using his or her built-in microphone and no headset, you will hear feedback. That is, you will hear yourself a couple of seconds after you speak because your voice is coming back through the other person's built-in microphone. Your Skype partner will think all is okay because he or she doesn't have the same feedback problem (you have eliminated echoes by virtue of using a headset). So, encourage people to use Skype with the proper equipment. Trying to talk through a feedback loop is no fun!

If you don't have a regular microphone, you can actually substitute it with any available set of headphones. Just plug the headphone into the micro-phone jack and speak into the earpiece. Of course, you need an extra headset to use as a, well, headset! Go ahead and try it. We did, and it actually works. We heard the giggles loud and clear.

Feedback you can see

Sometimes the person on the other end does not have a microphone handy or can't talk out loud because he or she is in a meeting, for example. In such a case, you may be the only one talking while your Skype partner is using the chat window. (See Chapter 6 for details of how to conduct a talk/chat simulta-neously.) This mode of communicating actually works more smoothly than you might think. Other uses for this alternative to voice chat is for skyping in a noisy office or with people with hearing difficulties. How many times have you wished that you could hold up a sign to a person on the other end of the telephone because the person couldn't make out what you were saying? The chat window is like having a built-in captioning service.

Making Your First Skype Call

Congratulations! Assuming that you've installed Skype, created a Skype Name, and tested your connections, you're ready to start skyping. The question is, Whom can you call?

Conducting a simple search for a fellow skyper

You can get started by searching for fellow skypers. To do that, follow these steps:

1. **Locate the Windows Start button on the lower-left corner of your desktop and choose Start➪Skype to launch Skype.**

   The Sign In to Skype window opens.

2. **Enter your Skype username (that is, your Skype Name) in the Skype Name box.**

3. **Enter your Password in the Password box.**

4. **Click Sign In.**

   The Skype program window opens. The menu across the top contains File, View, Contacts, Tools, Call, and Help. (Check out Chapter 3 for details about each of these menu items.)

5. **From the Skype menu, choose Contacts➪Search for Skype Users.**

   Alternatively, you can click the Search icon to bring up the Search window, shown in Figure 2-10.

6. **Enter the person's name in the uppermost text box.**

   In our example, the name Newbie Skyper appears in that box (refer to Figure 2-10).

   You can start your search by entering a full name. If your friend has not added his or her name to a profile, you may not be able to find the person this way. (The last section in this chapter provides some additional information about searching.)

7. **Click Search.**

   Skype searches through its directory to find the name you entered. When found, this name appears in the lower half of the Search window under "We found 1 contact."

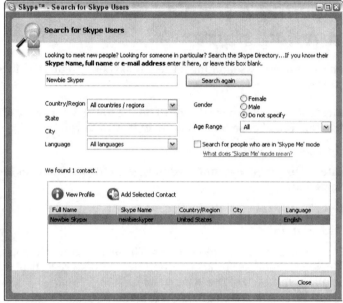

Initiating a call

Now that you've located someone you want to call, you're just one click away from a Skype conversation. To make a call from your Search window, follow these steps:

1. **Double-click the contact's Skype Name to bring up your contact's profile window.**

2. **Click the green call button.**

   This button appears as a green circle with a white telephone handset inside. When you click it, you hear the sound of a ringing telephone, and a call window appears. The call window contains:

   • The Skype Name of the person you are calling.

   • A photo, if posted, in his or her profile.

   • A red hang up button.

3. **When your contact answers the call, just talk!**

Sometimes your call is not accepted because you are not an authorized contact for that person. Receiving authorization and adding names to your Contacts list is an important part of communicating on Skype. This topic is covered at length in Chapter 4.

Using the Skype search directory

The ability to search Skype is a powerful feature. Although you can search for a Skype Name, full name, or email address, you can narrow your search based on a set of filters. An example of narrowing a search this way is shown in Figure 2-11. Although this search example is contrived, the search results are filtered using a specific age group and gender.

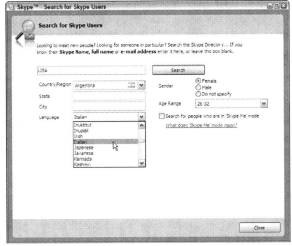

Figure 2-11:
Applying
filters to
restrict
search
parameters.

The Skype search directory gives you a number of ways to look for fellow skypers. Sometimes it is difficult to find the person you are looking for because he or she has not entered much information in a Skype profile. But you can still perform a successful search because you have several search options. If your Search window is not open, click the Search icon (looks like a spyglass) or, in the Skype menu, choose Contacts➪Search for Skype Users. You can search by adding one or more of the following search options:

✔ **Use the Search for Skype User box to:**

Enter a full name.

Enter an email address. This information is not revealed in a profile, but if you know someone's email address, you can use it to find that contact if the person put it in his or her profile.

Enter a Skype Name. Sometimes you have a Skype Name but no other information. You can search and then open a profile to find a time zone, full name, or Web site for further contact.

✔ **Search by location.** If you don't have a name or an email address, you can always try searching by location. If only a few Skype Names show up in an area, you might be able to find an individual. Looking by location also gives you an opportunity to meet skypers close by. To search this way:

Select a country or region from a drop-down list.

Enter a city, state, or both by entering the names into the boxes provided.

✔ **Search by language.** Select your language from a drop-down list. Language searches work best in combination with other parameters unless you are looking for something very specific, such as people who speak Vighur, Volapuk, or Shona living in Manhattan, Kansas.

✔ **Search by gender or age range.**

✔ **Search people in Skype Me mode.** Skype lets you display your online status (see Chapter 3), and Skype Me mode is an online status that lets the world know you are eager to receive calls and chats from anyone. When you search for skypers, you may want to select the Search People in Skype Me Mode box. Selecting this box guarantees that the people you find are people who want to talk.

You can mix and match all the search parameters. When you have everything entered in the Search window:

✔ Click Search to find skypers matching your search terms. Scroll through the results of your inquiry.

✔ Click Search again after changing the search parameters if you did not see the skypers you want to contact.

You can sort the order of the search results by clicking the appropriate header bar of the search results.

If you use multiple Skype accounts and have just one email account for all of them, all the accounts will turn up in a search.

Other than the email addresses you place into your profile, all the information you place into your profile is visible. People can find you by searching Skype using your email address only if they already know it.

Although you can opt to leave much of your profile blank, you will find that populating your profile with information such as language and country is helpful because this information is displayed in the search results. As the number of users in the global Skype community grows, people trying to find you have an easier time doing so with this extra information.

Chapter 3

Getting Familiar with Skype's Interface

In This Chapter

▷ Finding your way around Skype's menus and toolbar

▷ Announcing and changing your Skype status

▷ Getting help

So, now that Skype has found a home on your PC, you need to get in gear so that you can drive Skype with precision and ease. As with a new car, the console might be seem a little confusing at first. When you know where everything is, though, you'll find Skype to be a smooth ride. This chapter eases you into the driver's seat.

These pages familiarize you with the six main menus and toolbar in Skype so that you know right where to go for whatever task or activity you want to engage in. Consider this chapter as your tour guide for getting around in Skype. If we don't cover the full details of a given feature here, we tell you where in the book to go for more details about it.

You also find out about the various status icons that Skype uses to indicate a user's status (online, offline, and others) and how to change your status. Finally, we tell you about where to find the help you might need on occasion, steering you in the right direction both within the software and on Skype's Web site. The creators of Skype want you to go from 0 to 60 mph in the short-est amount of time so that you can feel comfortable and productive. The menu is designed to help you become a player in the Skype global commu-nity as quickly as you can say "Skype Me!"

Understanding the Skype Menus

Just as Skype is a breeze to install, Skype is equally easy to navigate and use. The Skype menus make doing essential tasks simple. Here's a quick summary before we get into the details: On the File menu, you establish your online presence, profile, and account information. The View menu enables you to control which elements you want visible in your main Skype window. On the Contacts menu, you manage the list of all your Skype buddies. Use the Tools menu to make conference calls, use SkypeOut, set up SkypeIn, send SMS messages, and open previous chats. The Tools menu also enables you to configure your Skype Options, or various settings. You use the Call menu to manage your calls. Finally, the Help menu assists you in finding specific information concerning setup information, checking for updates, and buying Skype credit. Read on for details of each Skype menu.

The File menu

Fasten your seatbelts! We begin our tour of the Skype menus with the options on the File menu.

Change Online Status

When you view your list of Skype contacts, you can immediately see whether your buddies are available to take a call. Every skyper, yourself included, has an online status. By default, Skype announces whether you're online and available to skypers who have you listed in their Skype Contacts list. You can override this default by changing your online status. Also, you get to control who can see whether you are online. (See Chapter 4 for details.)

Edit My Profile

Your Skype profile contains important information about you, some of which is used only for administrative purposes. Other portions of your profile are shared only with your Skype contacts. A third portion of your profile is always publicly available for any skyper to search. Use your Skype profile to set what information you want to appear in all three categories of your profile (see Chapter 4).

My Skype Account

There are two reasons to go to your Skype account: to revise your password and to manage your account. The latter includes tasks such as adding to your Skype credit (see Chapters 8 and 12) and setting up SkypeIn and Voicemail (see Chapter 8). You can review your history of SkypeOut calls, including to whom you made the calls, what dates and times, and amounts charged, if any.

Sign Out

When you are using Skype, you are logged in using a specific Skype user ID, called your Skype Name. Signing out doesn't actually quit the Skype program — it just logs you off and allows you to log back in using a different

Skype Name. Skype remembers which users you previously logged in as (see Figure 3-1). If you and several of your family members are sharing a computer, this feature makes it easy for one person to log out and another to log back in quickly.

When you are signed out, some menu options are still available to you. These include your Skype Account page, language settings, specialized network proxy settings (covered in Chapter 4), and the general help features.

Close

The close action simply closes your Skype window so that one less window takes up space view on your screen. Skype never stops running, and you can make and receive calls just the same. You just have a little less visual clutter.

Figure 3-1:
Skype
remembers
previous
Skype
Names used
when
logging in.

The View menu

The View menu displays which components of Skype's interface are visible on your screen (see Figure 3-2). Those that are visible appear with a check mark next to them. You can change the setting by selecting the item from the menu. Selecting a checked item in the menu makes the check mark disappear and causes the respective item in Skype to disappear. If you select an unchecked item in the menu, a check mark appears — and the corresponding item appears.

```
View
 ✔ View Toolbar
 ✔ View Addressbar
 ✔ View Statusbar
 ✔ View Dialpad
 ✔ View Live Tab

 ✔ Show Text on Toolbar
 ✔ Show Text on Tabs

   Show Contact Groups
   View Outlook Contacts

   Hide Contacts That Are    ▶
   Sort Contacts by Name
```

Figure 3-2:
A view of
your own.

Wisely, the Skype installation defaults make the toolbar, addressbar, and status-bar all visible (checked). You can remove these elements from view if you want, but why make life more confusing?

View Toolbar

The toolbar makes it easy to add and search for contacts, conveniently call ordinary phones and set up conference calls, start a text chat or send a short SMS message (see Chapter 10 for more info on SMS messages), send files from your computer to fellow skypers, and look up the profile of one of your contacts or people you are about to call.

View Addressbar

The addressbar is the location where you can type in a Skype contact name, a regular phone number, or even an Outlook contact name if you have Outlook running. You can find out more about your Outlook contacts later in this section.

If you have History Quickfiltering turned on (this is one of the advanced options settings discussed in Chapter 4), you need enter only a portion of the name or phone number. Those contacts or phone numbers with portions of text that match your entry are filtered and displayed. This handy feature enables you to locate your contacts easily and quickly.

View Statusbar

The left side of your statusbar shows your current online status. When you click the status icon, a context menu pops up and allows you to change the status from the available options. These options are spelled out a little later in the chapter.

While you're on a call, the statusbar displays the name of whom you are speaking to. If you're on a conference call, the statusbar displays that fact.

The right side of the statusbar displays how many skypers are online.

View Dial

Use the Skype Dial tab to assist you in making calls to ordinary phones. This facility is especially useful in placing international calls. Three quick steps are involved:

1. **Click the flag symbol or country name to which you are placing a call.**

   This action sets the appropriate country or region code.

2. **Enter the local number along with area code.**

3. **Click the green call button to place the call.**

There is also a link for calling rates.

If you are calling a place of business, you may be prompted to click certain keys such as the pound (#) or star (*) key, or an extension number. Click the dial pad keys as appropriate.

Unlike a PC-to-PC Skype call, SkypeOut and SkypeIn calls are not encrypted over the connection from the landline or cell phone up to the point where the phone line signal reaches the Internet. If you need to conduct a secure call, make sure that both you and your colleagues are connecting from PC to PC.

Show Live Tab

Those of you who are using the latest version of Skype may have the option to view ongoing Skypecasts directly from your SkypeLive panel. Selecting the Show Live Tab menu item makes the SkypeLive tab and panel visible.

Show Text on Toolbar

The icons in your Skype toolbar for adding contacts, searching, calling phones, conference calling, chats, sending SMS, files, and looking up a skyper's profile are pretty easy to figure out. Although you can forego text captions underneath each of the icons, you lose nothing by keeping the text captions.

Show Text on Tabs

Each of the tabs (including Contacts, Dial, and History) in the Skype menu has an icon as well as a brief text description. If you want, you can eliminate the text description appearing next to the icon on the tabs.

Show Contact Groups

As your list of contacts starts to grow, you may find organizing them into specific groups helpful. For example, perhaps you hold conference calls with members of a committee you chair in a company of organization. Try selecting all of them if your Contacts list grows from multiples of tens into the hundreds, or possibly thousands! In this case, you can definitely benefit from defining groups and use them to organize your contacts.

Show Outlook Contacts

One of the nice features of Skype is how you can easily tap into your list of contacts that reside in Microsoft Outlook. To make use of this capability, simply run Outlook at the same time that you have Skype running. Skype can read the list of Outlook contacts and phone numbers. When you look at your Contacts list within Skype, you see your Outlook contacts as well. The steps involved are as follows:

1. **Start Skype and then start Microsoft Outlook.**

2. **In your Skype application, if no check mark appears next to Show Outlook Contacts in the View menu, choose View⇨Show Outlook Contacts.**

   When you view your contacts, your Outlook contacts and their phone numbers are listed along with the Skype contacts. You can call anyone from your Outlook contacts directly from Skype.

Contacts menu

Another great feature of Skype is not having to worry about toting your list of contacts and installing them on each computer you run Skype on. Skype conveniently manages your Skype contacts on a central repository some-where out in the "Skype-o-sphere." The Skype Contacts menu (see Figure 3-3) has tasks for adding contacts, searching for Skype users, importing contacts, sending contacts to another skyper, backing up and restoring contacts to and from a file, and managing blocked users.

Figure 3-3:
Find all your
Skype
Name and
SkypeOut
contacts in
this menu.

Read on for useful information about the options in the Contacts menu.

Add a Contact

When you add a contact, you can do so by one of two methods (see Figure 3-4):

✔ **Find someone already in the Skype central database of Skype users.**

✔ **Add a SkypeOut phone number of someone who is not in any database:** For example, if you want to add your favorite cousin Jack to your list of contacts but Jack is not yet using Skype, you can list Jack's telephone number as a SkypeOut phone number. This way, you can quickly call Jack from your Skype contacts.

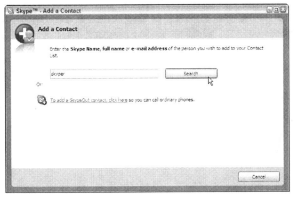

Figure 3-4:
Adding a
Skype user
or a new
SkypeOut
phone
number to
your
Contacts
list.

Search for Skype Users

Selecting this option opens a general search page that allows you to search the Skype central repository for registered Skype users (and there are more than 100 million of them!).

Here are several reasons you may want to search for a Skype user:

✔ **To call or chat with a person.**

✔ **To add that user to your Contacts list and ask for the user's contact details.**

✔ **To look up the user's Skype profile to find out about that person.** As a practical matter, this may be a person who is contacting you and requesting that you share your contact details. You may not recognize the person's identity. Looking at the person's public profile may help you figure out who is contacting you and determine his or her purpose for reaching you.

See Chapter 2 for more about searching for fellow skypers.

Import Contacts

Selecting the Import Contacts option launches a wizard that lets you import your contacts from Outlook and the Outlook Express Address Book. The import process does not require that you have Outlook or Outlook Express actively running.

If you're using Windows, follow these steps to import contacts from Outlook (steps for importing on the Mac are covered next):

1. **Select whether you want to import your contacts from Outlook, or the Outlook Express Address Book, or both.**

   When the import process is about to start, Microsoft Outlook pops up a window and tells you that another program is trying to access your email addresses stored in Outlook.

2. **Select the Allow Access check box and then set the duration of allowed access to, say, 10 minutes (see Figure 3-5).**

Figure 3-5:
Allowing
Skype to
access and
import your
Outlook
contacts
list.

While Skype is importing names from Outlook, it also searches the Skype central repository to see whether any of the people who are listed in your Outlook contacts list are already registered users on Skype. At the end of its search, it provides you a list of potential matches.

3. **Review the Skype profile of your potential matches and select the names you want included.**

   When Skype finds Outlook contacts to be imported that have email addresses and are not already on your Contacts list, Skype asks you to select any of those contacts you would like to invite to join Skype.

4. **Click Finish to complete the import process.**

You can always re-import your Outlook contacts without worrying that you will create duplicate entries in your Skype Contacts list.

Importing on the Mac works much the same as it does on Windows:

1. **On the Mac platform, choose Contacts⇨Import Contacts.**

   The Import Contacts window appears.

2. **Select the various data sources and type of information you want to import.**

When you import contacts with versions of Skype for Mac earlier than 1.5, be aware of the following caveats:

✓ You can import phone numbers only from the system Address Book.

✓ Telephone numbers appearing in the Address Book must begin with the international plus sign (+) prefix in order to be imported.

✓ The only information that gets imported is phone numbers. No other general contact information gets imported.

When you import contacts with version 1.5 of Skype for Mac:

✓ You can specify whether you want to import contacts from the system Address Book, Microsoft Entourage, or both.

✓ You can restrict the import to phone numbers only.

✓ You are not restricted to importing phone numbers starting with a plus sign (+) prefix.

Send Contacts

A nice and probably underused feature of Skype is the ability to forward contacts from your Contacts list to other Skype users. You can forward contact names to any of your Skype buddies.

When you forward Skype contact names, you have to pick and choose which contacts you are sending (that is, you are not obligated to send your whole contacts base but can send only the individual contacts you select for sending).

If the people listed in the contact names you are forwarding set their privacy settings (see Chapter 4) so that only authorized people can call or chat with them, your Skype buddies may need to request permission to communicate with them.

If you forward a SkypeOut contact to your Skype buddies, they will be able to immediately SkypeOut to that contact, just the same as if they were placing a regular SkypeOut call.

Backing up and restoring contacts

It's a good idea to back up your Skype contacts even though the list is maintained at Skype's central repository. Choose Contacts⇨Advanced⇨ Backup Contacts to File and select a location on your hard drive or USB flash drive to store the contact info. The file gets saved as a VCF file, which is a special format for exchanging contact info. If you're curious, you can open up the VCF file with a text editor such as Notepad.

Restoring your contacts is just as simple. Choose Contacts⇨Advanced⇨ Restore Contacts from File, locate your backup file on your drive, and click OK.

Manage Blocked Users

Anytime a person makes a person a nuisance of him- or herself, you have the option of banning this person from directly initiating a call or chat with you (choose Advanced⇨Manage Blocked Users). A person who is blocked cannot call you or initiate a chat with you, see your online status, view your picture, or do or anything else that your regular contacts have available.

Keep in mind that if you block a user, it will not prevent someone else from including both the blocked user and you on the same chat or conference call.

When you choose Advanced⇨Manage Blocked Users, a window appears that shows any currently blocked users and gives you the ability to take anyone off that list.

The Tools menu

The Tools menu (see Figure 3-6) provides commands for such tasks as creating conference calls and setting up SkypeIn, SkypeOut, and call forwarding. Within this menu you can also manage chats, set your language, and specify options.

Figure 3-6:
The Skype Tools menu. The little hand appearing to the left of a menu item indicates that this item connects to a Web page.

Create a Conference Call

Conference calls are a great way to huddle with a small group of people and work collaboratively. Skype conferencing is extremely easy to set up. Follow these steps:

1. **Choose Tools⇨Create a Conference Call or click the Conference Call button in your Skype toolbar.**

   A window opens in which to set up your conference call.

2. **To add participants, click the Add button or drag the contact name to the list box showing your conference participants.**

   You can also edit the conference title.

3. **Click OK when you are ready to start the conference.**

SkypeOut

Voice over IP (VoIP) is about using the Internet as the transport medium to make calls from one person's computer to another. This is fine, but Skype does it one better: A feature called SkypeOut lets you connect and make calls from your computer to a regular landline or cell phone. Pretty cool.

Depending on where you are calling to (and from), SkypeOut can be totally free or there may be a small fee for its use. If you want to place a call using SkypeOut, you may need to purchase Skype credit. Choosing Tools⇨ SkypeOut lets you purchase Skype credit if you need it (see Chapter 8).

Incidentally, when you hold conference calls, the participants on the calls can be speaking from regular landlines. You can find out the details and plenty more about SkypeOut in Chapter 8.

SkypeIn

The flip side of SkypeOut is SkypeIn. Anyone can call you from a regular or cell phone and reach you on Skype. All that's involved is setting up a SkypeIn phone number. Choose Tools⇨SkypeIn to set up a SkypeIn number. If you want to establish an international presence for your business, or want to make it easy for your aunt who lives overseas to call you from a local number in her country, SkypeIn is definitely great news. There is a fee for setting up a phone number, but it is low. Get the scoop on SkypeIn in Chapter 8.

Voicemail

You can't sit by your computer 24 hours a day waiting to receive incoming calls over Skype anymore than you can be near your home telephone all the time. Skype provides an optional Voicemail service that makes it easy for people to leave you messages when you are unavailable to take their call. Choosing Tools⇨Voicemail opens the Call Forwarding and Voicemail panel, where you can record an audio greeting message as well as set your Voicemail preferences. You can explore the details about Voicemail in Chapter 10.

Call Forwarding

One of the convenient features of Skype is the ability to automatically forward incoming calls. The call can be forwarded to a Skype Name of your choosing. Optionally, it can be forwarded to a regular telephone number, although you may have to pay a fee for SkypeOut charges. What happens if the Skype number you are forwarding to doesn't pick up the call? Skype allows you to designate up to three call forwarding numbers, and each is tried in succession.

To use Call Forwarding, follow these steps:

1. **Choose Tools⇨Call Forwarding to open a window and choose your call forwarding settings.**

2. **Select the Forward Calls When I'm Not on Skype check box and enter a phone number.**

   If you need more than one forwarding number, click the <u>Advanced Settings</u> link. (See Chapter 9 for more details about Call Forwarding.)

Send SMS Message

Skype has a Short Message Service (SMS) through which you can send brief text messages to mobile phones and devices that support the Global System for Mobile Communications (GSM). To use this service, follow these steps:

1. **Choose Tools⇨Send SMS Message.**

2. **In the Add to SMS window, select names appearing in your mobile-enabled and SkypeOut contacts and click the Add button to add them to the SMS Recipients list box.**

   Alternatively, just click and drag the name over to the SMS Recipients list box.

   If you want to send an SMS message to a phone number that's not on any of your lists, space is available for entering the number. Click the Add button to add it to your list of recipients.

3. **Click the OK button in the Add to SMS window and then enter your message in an SMS window; click the Send SMS button to send it.**

See Chapter 9 for more details on SMS messaging.

Ringtone

Skype lets you set the kind of ringtone you want to hear when someone calls you. Choosing Tools⇨Ringtones lets you navigate and select from items in a submenu. You can find out more about customizing ringtones in Chapter 4.

If you are using a wireless headset and someone calls you on Skype, it would be nice to know who's calling without having to look at your computer screen. There is third-party software that you can obtain to accomplish this; read about it in Chapter 9.

Share Skype with a Friend

To encourage your friends or others to try Skype, you may want to send an eCard with information about Skype and its download links. Choose Tools➪Share Skype with a Friend to open a form on which you can construct an invitation to use Skype and have the form emailed.

Recent Chats

Skype supports text chats between you and your fellow skypers. Not everybody will sit by his or her computer to keep a chat session going. And you may be in the habit of shutting down your computer every night before you go to sleep. What happens to your chat? Conveniently, Skype remembers the chat content and lets you resume where you left off.

To resume a chat, choose Tools➪Recent Chats and select from a list of recent chats. If the chat has not been given a title, what shows is the last line of the chat. (See Chapter 6 for much more about chats.)

Bookmarked Chats

Although the Recent Chats feature is useful, you may want to retain the chat to keep it live and ongoing indefinitely. Skype gives you this ability when you bookmark a chat. Choosing Tools➪Bookmarked Chats lets you open and resume your bookmarked chats. See Chapter 6 for more about bookmarking a chat.

Change Language

Parlez-vous français? Sprechen sie Deutsch? ¿Habla español? The Skype program can be set to more than 27 different languages. You can set your language preference by choosing Tools➪Change Language and picking one of the languages listed in the submenu.

Options

Skype lets you set your application preferences for the following features: General options, Privacy, Sounds, Sound devices, Hotkeys, Connection, Updates, and Advanced settings (see Chapter 4); Videoconferencing (see Chapter 7); and Call Forwarding, Voicemail, and SMS Messages (see Chapter 10).

The Call menu

The Call menu enables you to manage your calls. If the Call menu appears grayed out, you are not actively engaged in any calls on Skype. While you are actually on a Skype call, or receiving an incoming call, the menu is no longer grayed out.

The following options are available from the Call menu.

Answer

You can answer an incoming call from the menu, but it is probably easier to click the green call button that is displayed with incoming calls.

Ignore

This option lets you ignore an incoming call so that you won't hear it ring and can continue your activities without interruption.

Resume

You can place a call on hold; Resume enables you to resume the call.

Hold

You can place a call on hold and answer an incoming call if you happen to already be talking to someone on Skype and receive a call from someone else.

Mute/Unmute Microphone

During a call, you can select Mute Microphone. When you do, your microphone stops sending sound to the audio-in device. While the microphone is muted, the Mute microphone option changes to Unmute microphone.

Hang Up

At any time during a call, you can click the red hang up button to end the call.

All the items available from the Call menu are generally available to you by clicking various Skype buttons and toolbar icons. Unless you need to navigate the Call menu for these options, you may find clicking a visible button to be easier and more convenient.

The Help menu

New Skype users typically wonder about a variety of issues. What kind of headset should I get? How do I get Skype Credit? Why are my chats archived? To answer these and other questions, Skype provides a variety of online help facilities through the Help menu. Some of these answer questions; others take you to Web pages (see Figure 3-7) for specific tasks such as buying Skype credit.

Figure 3-7:
The www.
skype.
com Help
page offers
FAQs,
technical
support,
input from
the Skype
community,
and other
forms of aid.

Help

Choosing Help➪Help opens your Web browser to the www.support.skype.com site. On this Web page, you can find links for troubleshooting, user guides, announcements from Skype, and other kinds of helpful information. The Web page also contains a searchable a knowledge base. (By the way, you may want to check out Appendix B of this book for a variety of troubleshooting tips.)

FAQ

Choosing Help➪FAQ takes you to a page with answers to plenty of Frequently Asked Questions.

Getting Started

This option opens a sequence of screens to give you an ultra-quick tour of Skype. The very first time you install and launch Skype on your computer, this Getting Started guide is displayed. It continues to be automatically displayed every time you launch Skype until you select the Do Not Show This Guide at Startup option.

Get Headset

This option takes you to the Skype Web store, where you can purchase Skype-certified headsets, speakerphones, webcams, and the like. Chapter 11 profiles some of these products.

Buy Skype Credit

Certain features of Skype, such as SkypeIn and some ringtones, entail a fee. For this reason, you may need to buy Skype credit. You can find more information about Skype credit in Chapter 8 and credit for Business Groups in Chapter 12.

Redeem Voucher

Many of the Skype-certified gadgets on the Skype Web store as well as some of the products described in Chapter 11 come with voucher numbers for various Skype-related features, such as free SkypeOut minutes or Voicemail. Choose Help➪Redeem Voucher to claim your free services.

While we're on the subject of vouchers, look to the end of this book for vouchers to use toward purchasing various software products for Skype. Take advantage of these offers while they last!

Check for Update

Keeping up-to-date with the latest and greatest version of Skype might be a chore if you had to remember to check for updates. Luckily, Skype makes it easy for you to automatically keep your software up-to-date because Skype does the checking for you, as follows:

- ✔ Choosing Help➪Check for Update lets you check whether you have the latest version of Skype installed.
- ✔ Choosing Tools➪Options➪Updates lets you configure Skype to either alert you about new updates or make them automatically.

For major releases, you have the option to have Skype automatically download the new release, alert you that a new release is available and ask you for permission to download, or ignore new releases altogether.

Every so often, Skype may have a hotfix to a specific problem for a small number of Skype users. For these hotfixes, Skype gives you the option to automatically download, ask before downloading, or ignore updating for hotfixes altogether.

Report a Problem

If your Skype experience is not the way it's supposed to be, you may want or need to report it to Skype. Problems can span a broad spectrum of issues ranging from general support issues to incorrect charges in your Skype account.

When reporting a technical difficulty, you may enter a long and detailed message and click the Submit button, only to find that your message has not been submitted and you are instead presented with a laundry list of possible explanations for your problem. At the end of that list is a *true* Submit button. Skype wants you to be absolutely sure that your question is not responded to in its prepared list of responses. Do not be intimidated by this! Click Submit to have your issue addressed by some responsible person in Skype.

About

For troubleshooting purposes, it is helpful to know the version of Skype software you are running. Choosing Help⇨About displays this information for you.

Using the Skype toolbar

Although you can access many of Skype's features from the main menu, the Skype toolbar displays many of these essential features on toolbar icons. The base set of icons includes the following:

- **Add Contacts.**
- **Search:** Find registered Skype users.
- **Call Phones:** SkypeOut from the Dial tab.
- **Conference:** Start a Skype conference.
- **Chat:** Open a chat with the selected Skype contacts.
- **Send SMS:** Send an SMS text message to an SMS-enabled mobile device.
- **Send File:** Transfer a file to one or more Skype contacts.
- **Profile:** Look up the profile of a selected Skype contact.
- **Hold/Resume:** While you are actively talking to someone on Skype, this Hold icon appears. Choosing Hold holds the call and the icon changes from Hold to Resume. It continues to stay that way until you either click Resume to resume the call or the call is ended, possibly by the other person's hanging up. You don't want to keep someone waiting on hold forever, do you?

  When you are not actively on a call, the Hold and Resume toolbar icons are hidden altogether.

- **Mute/Unmute:** While you are on a call, you can mute your microphone. The icon changes from mute to unmute and stays that way until you unmute the connection or the call is ended.

  When you are not actively on a call, the Mute and Unmute toolbar icons are hidden altogether.

We need to mention some other important features of the Skype toolbar, as follows:

✔ If your Skype window is narrower than the space available to display the toolbar icons, some of them are hidden. In their place on the right side of the visible toolbar is a double-arrow icon. Click the icon to display the hidden toolbar icons and then click the icons that you need.

✔ As Skype continues to evolve, the specific toolbar icons that are generally available may differ from those shown in Figure 3-8.

Figure 3-8:
The Skype
toolbar.

Status Icons: Announcing and Changing Your Online Status

Your online status icon broadcasts your availability to the world. Changing a status icon from the Change Online Status menu is a piece of cake. However, knowing what each icon means may seem obvious at first, but the settings need explanation.

Displaying your Skype online status

Online status icons signify how you want the world to view you on Skype.

✔ **Offline:** Select this when you don't want to take or make any calls. The green call button for making calls is dimmed, indicating that you cannot make calls. However, you can look up profile information about any Skype contact while in this mode.

✔ **Online:** This is the status you have (the default setting) when you log on.

✔ **Skype Me:** This mode lets everyone else on Skype know that you are interested in talking or chatting with other folks even if they do not know you. By choosing this option, you temporarily disable your privacy settings. In this way, someone can reach you without having to request that you share your contact details. When you return from Skype Me status, your privacy settings are automatically restored.

✔ **Away or Not Available:** If your computer has been idle for a while, your status changes from Online to Away or Not Available automatically until you do something, such as move the mouse on your screen. Then it returns to Online status automatically.

- • An Away icon (a Skype icon with a small watch next to it) appears next to your Skype Name if there is no activity after a few minutes. Skype sets this time to five minutes, but you can change the settings to a different amount of time (see Chapter 4).

- • A Not Available icon (a Skype icon with a lunar crescent — a kind of "gone to sleep" mode) appears next to your Skype Name after your computer is idle for a longer period of time. Skype sets the amount to 20 minutes, but you can change the settings to a different amount of time (see Chapter 4).

The Away or Not Available mode tells other Skype users that you have Skype running but may have actually stepped away from the computer and are unable to respond to a call or chat. Actually, you could be snoozing at the keyboard, but nobody can tell the difference!

✔ **Do Not Disturb:** You can indicate that you don't want to be interrupted by choosing this status. In this mode, you won't receive any incoming calls or messages, but you can make outgoing calls.

✔ **Invisible:** This is sort of a stealth mode by which you can function as you regularly do in Online mode, making and receiving calls, participating on chats, and so on. When others on Skype look at your status, it appears that you are Offline, even though you're not.

You can change your status anytime, with a choice of ways to do that:

✔ **Choose File➪Change Online Status:** This method makes a drop-down list of status icons appear (see Figure 3-9).

✔ **Click the Status icon at the bottom-right corner of the window:** Clicking this icon brings up a drop-down list with Change Status as the first option. Click that option to open the same drop-down list that opens by clicking File➪Change Online Status. In Figure 3-10, the Online status icon is clicked to change the status to Away.

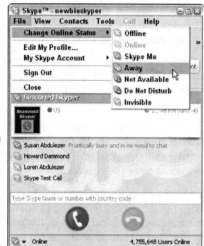

Figure 3-9:
Choosing
the status
icon from
the Skype
File menu.

Figure 3-10:
Choosing
the status
icon from
the system
tray.

Figure 3-11 shows what happens when you access the Change Online Status
icons from File⇨Change Online Status.

Figure 3-11:
Setting your
online
status by
clicking
the Skype
status
icon and
selecting
the desired
status.

Part II
As You Like It: Skype Your Way

The 5th Wave   By Rich Tennant

©RICHTENNANT

"Mona, this is no way to deal with your chat-line addiction."

In this part . . .

In this part, you discover how to fine-tune your Skype settings and personalize your profile. You also get the full scoop on online chatting through Skype, including how to use chats one on one and in a group setting, and save, bookmark, and organize your chats. We show you how to find that needle-in-a-haystack piece of information from a prior chat as well as how to share files instantly with other skypers.

This part also gets you set up with using video through Skype and shows you how to make calls out to regular old telephones from your computer using SkypeOut. You also find out how to receive telephone calls that you can answer through Skype on your computer — using a service called (as you can guess) SkypeIn.

Chapter 4

Customizing Skype Options to Suit Your Style

*W*hen you've started to get used to Skype, you may want to fine-tune some of the details regarding how it works in your digital world. The more you know about making your Skype settings suit your daily needs, the better Skype will serve you.

These pages guide you through choosing your general settings, privacy options, notifications and alerts, sound effects and sound devices, hotkeys, connection options, and advanced features. We cover these option groups in the same sequence as they appear in the menu pane of the General options menu. (We skip the call forwarding, voicemail, and video categories here, however, because those topics are covered in detail in Chapters 7, 9, and 10.)

Use this chapter to help you define how you interact with the Skype community and what you will (and won't) let the software do.

Fine-Tuning Your General Options

To get started, choose Tools⟳Options to open the Skype Options menu, which contains all the options covered in this chapter. You can see the option groups listed in the pane on the left side of the screen. Click the top one, General, to open the General menu (see Figure 4-1).

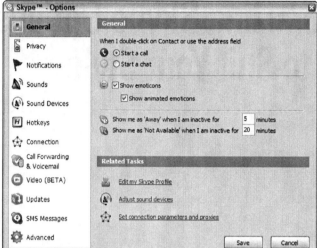

Figure 4-1:
Set your
General
settings by
choosing
Skype
Tools⇨
Options.

General options are grouped into three categories:

- Telling Skype to start a chat or start a call when you double-click a contact
- Emoticon behavior
- Online status timeouts

To speak or not to speak

Maybe you're a little shy when you first meet someone, or you just don't feel like talking sometimes. You can tell Skype that you always want to start in chat mode (which really means writing mode) rather than audio mode whenever you double-click the name of a person in your Contacts list (see Chapter 2 for more about contacts). To have Skype start in chat mode, click the Start a Chat button under the When I Double-Click on Contact or Use the Address Field option. In chat mode, you can exchange written messages with another user, send files from your hard drive, and paste links of Internet addresses from any Web browser.

The alternative is to set Skype to start in audio mode by clicking the Start a Call button (refer to Figure 4-1). In this mode, when you double-click the name of a contact, you hear a ringing tone just as you do when you dial a person on a regular telephone. Assuming that your recipient is available, you are in a direct conversation with the person you called.

Regardless of whether you set your General options to start a chat or start an audio call when you double-click a contact, you are always free to directly call the selected contact and click the green call button. You are also free to start a chat with that selected contact.

Express yourself

Emoticons, an array of symbols that express emotions, play an important role in a Skype chat (see Figure 4-2). Meeting a new person over Skype by using the chat mode means that you see only the written words, not the person's facial expressions. You can't hear the eagerness, sincerity, kindness, or humor in his or her voice. Sometimes a chat mode is an introductory meeting, which, if all goes well, leads to a conversation, a file transfer, or other privileges we extend to people we trust. The addition of an emoticon can break the ice between new Skype contacts.

Although Skype emoticons are generally animated, such as a "(dance)" emoticon that displays an icon-sized person dancing, you can change the behavior of your emoticons to be static images in your chat windows. To do so, deselect the Show Animated Emoticons option in your General options. If you want to eliminate the graphic representation of emoticons altogether and instead show text that represents the emoticon, deselect the Show Emoticons option. The dancing person emoticon appears as "(dance)," with the parentheses, in your chat window.

Figure 4-2:
Express
yourself
through
Skype's
emoticon
gallery.

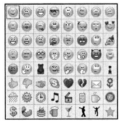

In addition to placing emoticons by clicking icons in the pop-up panel shown in Figure 4-2, you can enter the text representation in your text chat window. For example, to show the image of the emoticon with sunglasses, you enter **(cool)**.

Some undocumented emoticons that you might enjoy are also available. One of our favorites, especially after laborious hours of writing and editing, is the "(headbang)" emoticon. When you see this animation, you will know that emoticons can be very expressive. (Try finding some you like at `http://share.skype.com/sites/en/2006/05/new_emoticons_in_skype_25.html`.)

In your chat window, you can display emoticons representing the flags for various countries. To display a flag for the United States, enter **(flag:us)**; for Canada, enter **(flag:ca)**. The choice of country flags is extensive and is based on the two-letter designation set by the International Standards Organization standard ISO-3166. You can find the full list of two-letter country abbreviations at

```
www.iso.org/iso/en/prods-services/iso3166ma/02iso-3166-code-lists/index.html
```

Falling asleep at the keyboard? Set Skype to cover for you while you snooze

The other choices on the General options relate to automatic changes to your online status — your visibility to others on the Skype network. Click the box next to Show Me As 'Away' When I Am Inactive For to set how long you want Skype to wait before automatically signifying that you've stepped away from the computer. When your mouse movement halts, Skype senses this mouse inactivity after the amount of time you set and automatically displays an Away icon next to your contact name. You control how long Skype waits to trigger a change in status; you may choose to have an Away sign posted after five minutes, and a Not Available sign posted after 20 minutes. To change these settings, choose Tools⇨Options, click General, and adjust the number of minutes listed; then, click Save.

Protecting Your Privacy

Skype gives you two mechanisms for protecting your privacy:

- ✔ **Your Skype profile:** Information about yourself that you allow anyone on Skype to see. It can be jam-packed with information or contain barely anything.

- ✔ **Your Privacy setting:** This setting enables you to set who can directly call or chat with you.

In real life, anyone can find out where you live. All anyone needs to do is to look you up in a phone book, buy a mailing list, or find you through some publicly available information source. The fact that people happen to know you means that they can show up at your doorstep whenever they please.

But if you live in a building with a doorman, visitors have to be announced, and you can give instructions to the doorman as to who is allowed to visit and who isn't. Skype, through its profile and privacy settings, parallels this latter arrangement closely. The next two sections show how you can make it easy or difficult for people to find you based on your Skype profile, and how to assert control over who can start a call or chat with you based on your privacy settings.

Editing your Skype profile

Your Skype profile contains three kinds of information:

- ✔ **Information that all Skype users can see:** Includes information such as

  - **Your full name.**

  - **Your vital statistics:** Gender and date of birth.

  - **Geographic location:** You can specify the Country/Region, State/Province, and City.

  - **Your preferred language.**

  - **Your telephone numbers:** Can include home, office, and mobile.

  - **Your Home page:** You may have a Web site that you want the world to visit.

  - **A description about yourself:** If you want to tell your life's story, you must do it in 200 characters or fewer.

- ✔ **Information that only your Skype contacts can see:** Only your Skype buddies get the exclusive and revealing picture of you, as well as some limited information about your Skype account.

  - **Your Skype picture.**

  - **Your time:** It can be your computer's local time or a Greenwich Mean Time (GMT).

  - **The number of Skype contacts you have:** You can choose not to reveal this information to anyone.

- ✔ **Information known only to you:**

  - **Your email address:** You can list more than one email address.

The information in the public portion of your profile is available literally to any Skype user. The semi-private information is reserved for your Skype contacts, and no one gets to directly view your email address.

From a privacy perspective, a few points are worth mentioning:

- ✔ Even if you don't want to reveal much information, you'll probably find that listing your preferred language and country is useful. Adding identifying information such as a country can be helpful to someone who searches for your name and finds several people with the same first and last name as yours. Additional information can help to narrow the list.

- ✔ Many people are unaware that the number of Skype contacts a person has is listed unless they deselect this option. You may not want the world to know that you have only three contacts, especially if you portray yourself as being experienced in VoIP and Skype.

✔ You can choose to fill in all the items in your profile, or none of them. You should have your email address in your profile so that you can take advantage of features such as SkypeOut, SkypeIn, and Voicemail. Also, if you ever forget your password, you'll need your email account to reset your password.

✔ Although your password is never directly displayed, a Skype user who generally knows your email address can find you by searching Skype.

The last item may prove to be very helpful. You can use, for example, a sales@*company*.com or info@*company*.com email address for your company. People can then easily search for and find the right party, such as sales or support, if your company has many Skype user accounts.

Establishing who can call or chat with you

Privacy options let you allow or deny calls and chats. You can also put some callers in the virtual doghouse by blocking contact altogether. Be aware that privacy is something you must choose. Skype's default setting allows anyone to initiate a call or chat with you. Skype calls this process "Authorization," meaning that when someone requests your Contact Details, you have authorized him or her to know both your full Skype identity particulars (except your email address) and your online status in real time. Authorization is usually what you want to choose in order to keep connected to your new circle of friends. This setting is probably not your first choice, however, when it comes to the world at large.

The Skype community is large and growing daily. This means that possibly thousands of people, organizations, or interest groups might want to contact you — or you them. Although this is a welcome opportunity, you need to be aware that you may be sharing a bit more than "hello" when you share contact details or accept a call or chat.

When a person contacts you for the first time, a Skype window pops up on your screen (see Figure 4-3). You might be tempted to hastily click OK so that you can get back to what you were doing. Wait! Before you prematurely send all your information, you can click the Show Options button and look over your other choices (see Figure 4-4), as follows:

✔ **Send Your Contact Details with This Person:** Choosing this option allows the person to freely communicate with you whenever he or she wants.

✔ **Do Not Send Your Contact Details with This Person:** Choosing this option ignores the person's request. That person can't tell whether you simply didn't respond or actually declined his or her request.

✔ **Block This Person from Contacting You in the Future:** This option means that you will *always* appear as Offline anytime that person views you in his or her Contacts list.

By deselecting the Add This Person to My Contacts option, you can share your contact details with a person without adding that person to your own list of Skype contacts.

Figure 4-3:
When requests pop up, your options are not immediately displayed! Click Show Options before you share your information.

Figure 4-4:
Options available when deciding to add a contact.

Even if you set your privacy settings so that only authorized people can contact you or see when you are online, you can still receive "Hello from" requests from new visitors.

Receiving Notifications when Someone Skypes You

Skype provides a number of notifications that alert you, by on-screen messages or through sound effects, when someone on your Skype Contacts list comes online or does the following:

- ✔ Calls you
- ✔ Starts a chat with you
- ✔ Sends you a file
- ✔ Requests your contact details
- ✔ Sends you contacts
- ✔ Leaves you a Voicemail message

To choose what notifications you want Skype to provide you with, click Notifications in the Options window to display the Notifications panel, shown in Figure 4-5. Depending on how you like to work, any or all of these notifications can serve you well or be an annoyance. You might want to start by enabling all the notifications (do so by clicking the box next to each option) and then see how they work out for a while. You can go back and alter your settings anytime.

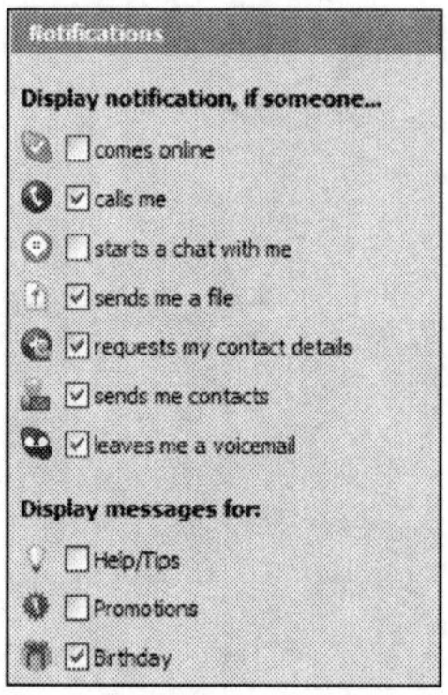

Figure 4-5:
Tell Skype
what types
of communi-
cation to
alert you to.

Another type of notification setting, which appears at the bottom of the Notifications menu, is Display Messages for Help/Tips. Enable this setting if you want Skype to alert you to various kinds of supplementary services available from Skype such as SkypeOut, SkypeIn, or sending SMS messages, which are fee-based services. In case you are wondering what an SMS message is, look at Chapter 9.

Skype can notify you when any of your Skype contacts come online. If you start accumulating a lot of contacts, you may find this feature distracting and sometimes even downright annoying. If you happen to use a wireless network and your connection cuts in and out, then after the connection resumes, you can get alerts displaying the contacts that appear to come online all at one time (see Figure 4-6). Sometimes, too much of a good thing may not be so good! For this reason, many skypers choose to keep the setting for Someone Comes Online disabled.

Figure 4-6:
Stacked
alerts can
unexpect-
edly clutter
your screen.

As you become familiar with Skype and don't require all the helpful hints and suggestions, you may want to turn off Help/Tips because it, too, is distracting by constantly suggesting and displaying links to services that have nothing to do with what you are doing at the moment, such as SkypeOut, SkypeIn, or sending SMS text messages.

Customizing Your Sounds

Remember the uproar around Classic Coke? The Coca-Cola Company changed the formula, introduced a new beverage, and suddenly everyone became nostalgic for the old taste. Well, the designers of Skype wisely kept its classic, familiar ringtones when they added a snazzy new group of modern sounds for the following:

✔ Busy Signal

✔ Call on Hold

✔ Connecting Call

- ✔ Contact Online
- ✔ Dial Tone
- ✔ Hang Up
- ✔ Incoming Chat
- ✔ Resuming Call
- ✔ Ringtone

Skype allows you to customize your ringtones by importing WAV files purchased on the Internet or even by importing your own digital recordings. Skype further lets you manage its behavior by controlling the sound devices on your computer. The following sections explain how to customize Skype in this regard.

Playing with bells and whistles

A virtual cacophony of sounds awaits your selection, depending on your mood and how much idle time you have to fool around and experiment with what suits your fancy. You can choose from a variety of dial tones, busy signals, connecting buzzes, on-hold tones, and many others from the sounds panel list (see Figure 4-7).

Figure 4-7:
Jazz up your
Ringtones
specifi-
cations.

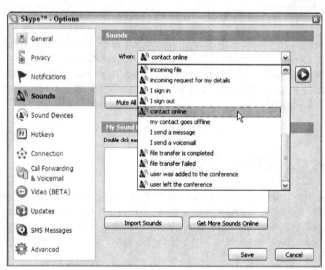

After you pick a sound type, you can choose the modern or classic variety.
If you wish, you can import sound files directly from your hard drive or
even purchase and download sound files from the Skype Web site. Choosing
Tools⇨Ringtones⇨Get Latest Ringtones takes you to Skype's personalization
Web site (see Figure 4-8).

Figure 4-8:
You can
obtain
musical
ringtones
from
Skype's
online
personaliza-
tion page.

This site allows you to try different ringtones before you choose to buy them
(see Figure 4-9).

Figure 4-9:
A Hip
Mozart
sound
sample.

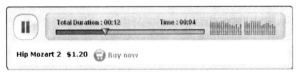

Skype can play WAV sound files. If you have an MP3 file, you can convert it
to a WAV-formatted file by using programs such as Audacity and QuickTime
Pro. Plenty of other programs make this conversion as well. Catch the latest
posts about this and other topics on our Web site for this book, www.
skype4dummies.com.

Changing your Sound Devices settings

Sound Devices settings pertain to Audio In, Audio Out, and Ringing. Choices here relate to Windows Default Sound Device and any custom Sound Cards installed on your computer (see Figure 4-10). Unless you have some special requirement, you should leave these options alone. If you're not sure whether your system has a special requirement, it probably doesn't.

Figure 4-10:
Sound
device
settings.

Sound Devices	
Audio In	Windows default device
Audio Out	Windows default device
Ringing	Windows default device
	☐ Ring PC speaker
☑ Let Skype adjust my sound device settings	

One situation in which sound device settings may need to be adjusted is when you wish to use a USB phone and speakers for a new HiFi system at the same time. In such a situation, you can do the following:

1. **Choose Start➪Settings➪Control Panel➪Sounds and Audio Devices.**

   The Sounds and Audio Devices Properties window appears.

   (Depending on your operating system and how you have it configured, you may have to navigate to the Windows Sounds and Audio Devices control panel differently from the way described in this step.)

2. **Click the Audio tab of the Sounds and Audio Devices Properties window and then select your sound card as the Sound playback default device.**

3. **Click the Voice tab and select your sound card as the Sound playback default device.**

   Click the OK button to save and close your Sounds and Audio Devices Properties window.

4. **In your Skype application, choose Tools➪Options➪Sound Devices.**

   The options to choose here are for Audio In and Audio Out. Choose your USB device for each.

If your voice sounds very faint to your callers, deselect the box next to Let Skype Adjust My Sound Device Settings; then, set your sound settings at the Windows Sound control panel.

Customizing Hotkeys

Keyboard jockeys are in for a treat. You can easily create your own hotkey settings to juggle your Skype tasks such as answering calls, putting someone on hold, hanging up, and more, all at the touch of a keystroke.

The following example makes clear how to create your own hotkey.

Suppose that you want to make the Skype window appear instantly at a keystroke. Here's what you can do:

1. **Click the Hotkeys menu appearing in the left panel of the Options window.**

   The items and buttons appearing under the Enable Global Hotkeys check box are all grayed out.

2. **Select the Enable Global Hotkeys check box to activate the hotkeys.**

3. **Select Focus Skype and then click the Change Hotkey button.**

   The window shown on the right side of Figure 4-11 pops up.

4. **Enter the letter A or any letter of your choosing in the text box and click OK.**

   This step enables the Focus Skype action.

5. **Save your Hotkey settings.**

Figure 4-11:
Setting a
hotkey.

What is Focus Skype and what is the WinKey?

Your computer can "focus" on only one application or window at a time. When you click in a window, your computer switches its focus to the program associated with that window. *Focus Skype* is a directive that tells the computer you want the main Skype window to appear. It's kind of like summoning the genie in the bottle, but the genie is Skype.

On the keyboard of many Windows-based computers is a special key that fires up the Windows Start menu. The key, called the WinKey, looks somewhat like a wavy flag. Because the key is not generally used for anything other than summoning the Windows Start menu, it's an ideal candidate for customized hotkeys in your applications.

Now you can instantly jump to Skype from any program in Windows whenever you press WinKey+A. If Skype is already running, the Skype client window comes to the foreground. If Skype is not running when you press WinKey+A, the Skype application launches as it normally does when you start Skype. If you don't have Skype configured for automatic logon, you need to sign in as you regularly do.

Connecting to the Internet when Skype Default Settings Aren't Enough

Connection settings relate to ports and proxy connections. Skype works over the same connections that you use when you surf the Internet with your browser. If you can connect to the Internet with your browser, and your Skype contacts can, too, you should be able to carry on a conversation.

Customizing your proxy settings

Some companies require you to apply special settings to connect your Web browser to the Internet. Generally, your Skype Connection settings should closely match those of your browser.

 Most of the time, you won't have to tweak your settings because Skype usually works without any fiddling.

If you require a custom connection, perform the following steps:

1. Choose Tools⇨Options⇨Connection from the Skype application menu.

The Connection panel appears.

Working with internal and external firewalls

A firewall watches incoming and outgoing network information and acts as a kind of traffic cop to halt or prevent anything that is not specifically allowed to slip through a defined defense perimeter. Typically, large corporations have dedicated firewalls that create a defense perimeter around the whole corporate network or specific network segments. These are external firewalls. If your company has an external firewall that intercepts and blocks incoming or outgoing Skype traffic, you will have to ask your network or security administrator to open the firewall to allow Skype traffic to move through the company network.

Many small companies and individuals do not have a dedicated external firewall, but instead have an internal firewall that runs in software on each individual computer. This kind of firewall sets up the defense perimeter around your computer.

2. **Set your mode for proxy detection and fill in host and port information.**

A drop-down list provides for Automatic proxy detection, HTTPS, or SOCKS. Your System or Network Administrator should be able to tell you whether your connection is going through HTTPS or SOCKS, and he or she should be able to give you the hostname (it may be an IP address) of your proxy server and a port number. Fill in this information.

3. **Enable proxy authentication.**

You may need to select the box for Enable Proxy Authentication and provide a username and password.

Your Skype proxy and connection setting should more or less parallel your Web browser connection setting.

Customizing your firewall settings

Remarkably, Skype works in many corporate environments without any customized setup even when your organization's firewall would stop other software programs from connecting to the Internet. For the most part, if your company's firewall allows you to connect to the public Internet through your Web browser, you should be able to use Skype without difficulty.

Here are some troubleshooting steps you can try if your firewall is stopping incoming or outgoing Skype traffic:

✔ **Run your Skype incoming connections through ports 80 and 443:** If you are having difficulty connecting to the Internet over Skype after establishing your proxy settings, look to see that a check mark appears next to Use Port 80 and 443 as Alternatives for Incoming Connections.

You can find this setting by choosing Tools⇨Options and clicking Connection from the list of options in the left panel. If the Use Port 80 and 443 as Alternatives for Incoming Connections option is not checked, select it, save your settings, and retry Skype.

✔ **Open your external firewall:** If your company is running an external firewall, ask your network or security administrator to configure the firewall to allow for Skype traffic.

✔ **Open your internal firewall:** If you are running your own personal firewall and have administrative rights on your computer, here are some steps you can take:

  • Run your personal firewall software.

  • Check to see whether there is currently a rule that specifically prohibits Skype traffic. The first time you ran Skype, your personal firewall server may have prompted you when it detected Skype traffic and suggested that it be blocked. If you gave the OK to block the traffic, Skype cannot communicate with the outside world. Your solution is to find the offending rule and remove it.

    If you can't find the rule that blocks Skype, the following may work: Uninstall Skype and re-install it. The next time Skype runs, your personal firewall may prompt you about what to do. This time, you can tell it to allow the Skype traffic.

    Your personal firewall software may not have any rules about Skype and may need to be explicitly told that network traffic associated with Skype should be allowed. You must add Skype to your firewall's list of allowed programs.

✔ **Find further information on the Skype Web site:** Go to www.skype. com/help/guides/firewall.html to find information about specific Skype firewall setups including Windows XP SP2 Firewall, Norton Personal Firewall, Zone Alarm Pro, and McAfee Firewall Pro.

Keeping Up-to-Date

Some people like having their application software always up-to-date. Others want to be told when a new version of their software is available, whereas others blissfully chug along without annoying reminders. Which kind of person are you? Whatever your preference, Skype lets you set it (see Figure 4-12).

Skype lets you set your preferences for both major updates by choosing Options⇨Updates and then selecting one of the following;

✔ **Download Automatically:** This option automatically checks to see whether a new version of Skype is available whenever Skype starts up; if one is, Skype automatically downloads it.

✔ **Ask before Downloading:** This option is similar to the automatic download, but it politely asks you for permission before downloading.

✔ **Ignore:** In choosing this option, you are instructing Skype to take no action to download new versions of Skype unless you manually choose to download Skype.

Between major releases, Skype may introduce minor releases or hotfixes. You have the same set of options available to you for downloading hotfixes as you do for downloading major releases of Skype. On the same screen where you set your preference for major Skype releases, you can set your preference for hotfixes and select Download Automatically, Ask before Downloading, or Ignore.

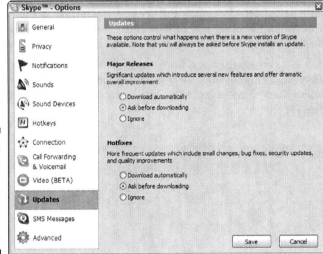

Figure 4-12:
Tell Skype
how you
want it to
behave
concerning
updates.

Improving Your Skyping with Advanced Options

Advanced options are not absolutely essential to using Skype, but they can make things smoother. The options are as follows:

✔ **Startup:** If you anticipate using Skype frequently, you may want to have Skype start up automatically whenever you power on your computer.

✔ **Automatically Answer Incoming Calls:** Unless you have a special need to automatically connect to whomever calls you at the very first ring, you should leave this option disabled. There are special circumstances when you would want to automatically answer incoming calls, however. One of them (setting up a "NannyCam") is covered in Chapter 7.

✓ **Enable Contacts List and History Quickfiltering:** Choosing this option tells Skype to quickly locate any contact in either your Contacts list or history of chats by simply typing in any portion of a name. Only those items that match appear. As your list of Skype contacts and chats grow, Quickfiltering becomes an indispensable tool.

✓ **Automatically Pause Winamp:** Winamp is a freeware player that enables you to play music and video content. Winamp has many plugins (see www.winamp.com/plugins) that enable you to use it in conjunction with other programs. The advanced settings in Skype (see Figure 4-13) let you automatically pause Winamp during an incoming call.

Suppose you're working at your desktop and enjoying some peaceful music on your computer's CD player. Suddenly you're interrupted by a Skype caller. You answer the call, but the music hasn't stopped playing! You're hearing both the music and your Skype buddy at the same time. Your buddy isn't hearing any of your music and doesn't know why you sound so confused as you frantically scramble to stop the music.

Wouldn't it be nice if the music automatically shut off when you answered your Skype call? If your computer uses Winamp to play music, as many computers do, you can enable the Automatically Pause Winamp During Calls feature.

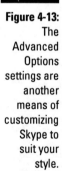

Figure 4-13: The Advanced Options settings are another means of customizing Skype to suit your style.

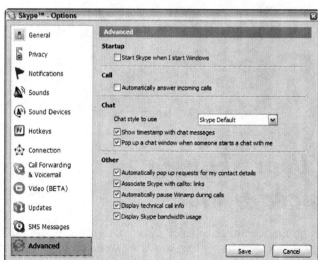

See Chapter 6 for Advanced Options settings pertaining to chats.

Chapter 5

Getting Personal

*W*e all like to make our mark, stand out in a crowd, and establish a unique identity. Broadcasting your personality to the world on Skype is not only fun but also may be a useful strategy if you use Skype as a business communication tool. Either way, you can find choices within the Skype program as well as third-party add-ons that allow you to strut your stuff. Personal choices include what logon name to choose, how much to say about yourself (covered in Chapter 2), and whether to post a photo, logo, or an avatar to show your face to the world. Decisions, decisions, decisions. It's up to you to make them, but it's up to this chapter to describe your choices and how to set them in motion.

Personalizing Your Skype Profile with Graphics

If you don't select a photo or image, Skype automatically inserts a cartoon-like "ball headed" icon for you. If you don't have a photo or artwork, you can use one of the ready-to-use icons that Skype provides (see Figure 5-1). When you click the Change button in your profile, the My Pictures window appears.

If you are creating your own graphic, you can use JPG (.jpg) files or bitmap (.bmp) files. The ready-made images installed with the Skype software use PNG (.png) files, so PNG files are also acceptable.

When you use an image, Skype automatically scales your picture to fit inside a 96-x-96–pixel swatch.

Figure 5-1:
Sample
swatch
of Skype
pictures.

All the text content of your profile is stored in a centralized Skype database; however, your picture is kept only on your local computer. If you are using Skype both on a computer at work and at home, you need to keep your picture file on both. You also need to adjust your Skype profile on each machine so that the same picture appears.

Having Fun with Avatars

Why limit yourself to a static picture if you don't have to? Skype allows you to create many ways to represent yourself visually. Among these are images called avatars.

The Hindu God Vishnu loved to change shape and appear in the form of a human or an animal. These forms are also called avatars. Because the personality is the same but the form is different, *avatar* became a handy word to describe images representing (and being controlled by) real people in a virtual environment.

However, computer avatars are not limited to human or animal forms. Some avatars are cartoon characters, robots, talking trees, teapots, smiley emoticons, and whatever your imagination can invent.

Skypers have taken avatar creation to a new level by making use of various avatar software tools. Your new Skype contact may come to you in the form of an image that moves, talks, expresses emotion, and generally reflects, in real time, what the "hidden" personality is saying and doing. You may find yourself talking to a little character that has been custom built to have green hair, violet eyes, a ten-gallon hat, and a golf club. (Sometimes our avatars say more about us than our real looks!) You may even see a dog talking while hearing your friend's voice in a Skype conversation.

Oddly, the avatar concept has come full circle. You not only can appear in the form of dog, cat, or coffee cup but also can take your own photograph — your real image — and turn that into a talking avatar. So, the real you takes on your own avatar form!

Creating WeeMees as avatars

A very simple and inexpensive type of avatar you can create is called a WeeMee (see Figure 5-2).

When you click the Create My Avatar button, you are instantly whisked away to an avatar construction factory (see Figure 5-3).

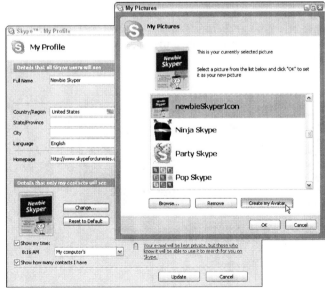

Figure 5-2: Click the Create My Avatar button in Skype to start creating a simple avatar.

Figure 5-3:
Web
construction
site for
building a
WeeMee.

You can try your hand at building a WeeMee by following these steps:

1. Click the Build Your WeeMee Now! button.

You are presented with a WeeMee without features and holding a sign saying Male/Female.

2. Click the gender you want.

All the options for the WeeMee of that gender appear. Depending on the options you select, you may have some animations in your avatar. For example, the boat and clouds of the WeeMee in Figure 5-4 are moving. (It's hard to make out in black and white here, but the boat is to the left, chugging down the waterway.)

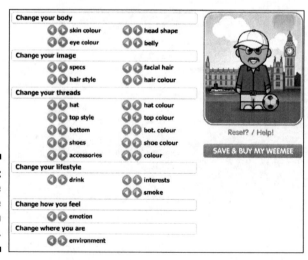

Figure 5-4:
The Skype
WeeMee
construction
kit.

You have a wide range of options for your WeeMee, including hair styles, fashions and interests. Keep exploring the offerings until you're done creating your WeeMee; then, click Save and Buy My WeeMee.

Building animated avatars with CrazyTalk

Are you ready to let your creative streak run free? Animated avatars that you can create with CrazyTalk Messenger take image interactivity to the next level. CrazyTalk comes in two versions: a free version called CrazyTalk for Skype (go to www.reallusion.com/crazytalk4skype) and a paid version called CrazyTalk Messenger (go to www.reallusion.com/crazytalk). The free version comes with a fixed selection of avatars that you can use but can't modify. CrazyTalk Messenger allows you to design and build avatars that you also can use in Skype. You can use photos of yourself. And by the way, having a photorealistic avatar of yourself on Skype may be the perfect solution for when you're having a bad hair day!

What do you want to make your animated avatar do? Maybe you want your character's mouth to lip-synch with yours as you speak on Skype, or perhaps you'd like it to capture a mood to suit the occasion.

Building avatars with CrazyTalk Messenger is surprisingly easy. We don't have the space here to step you through all the features of CrazyTalk, but we can tell you enough to whet your appetite. The first step is to import an image of the character you want to work with. The image can be a photo, a cartoon, or a scan of a portrait, for example. As long as it is a standard graphic such as a JPG file, that's fine. You can find some free and mostly copyright-friendly images at the following sites:

```
http://www.pics4learning.com
http://www.freephotos.se
http://en.wikipedia.org/wiki/Public_domain_image_resources
```

The best topics to search for are animals, portraits, and cartoons. Although you can find a wealth of images to choose from, be aware that some images may have copyright notices and usage restrictions.

To import a picture into CrazyTalk, follow these steps:

1. **Click the top icon on the vertical toolbar on the left side of the Model Page.**

    The Open File window opens, enabling you to browse for image files on your hard drive. (Incidentally, CrazyTalk supports directly importing an image of a printed photo from a scanner attached to your computer.)

2. **Navigate to the JPEG or BMP image file you want to import, click it, and then click the Open button.**

There's more to animating an avatar than just importing a picture. When the avatar's jaws move in sync with your speech and its eyes blink, you want it to appear natural. You also need to draw some simple contour lines so that CrazyTalk knows how to smoothly morph the avatar's face when it is speaking or moving its head slightly.

After you import the picture, you can rotate the image so that the avatar's face is upright and vertical. If the face is slanted, the jaws can't move naturally. To fix the image, identify the spindle points for the lips and eyes (see Figure 5-5), as follows:

1. **Click the Auto-Fit Anchor Points toolbar icon (the fourth icon in the vertical toolbar at the left).**

The Auto-Fit Anchor Points window, shown in Figure 5-5, opens.

2. **Drag the points over your imported image to match the reference image on the right; then, click the OK button.**

CrazyTalk processes the image and generates a set of contour lines.

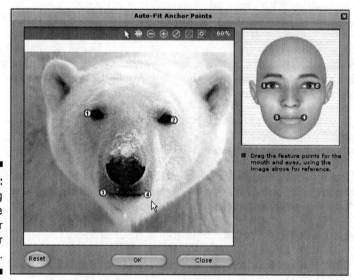

Figure 5-5:
Adjusting spindle points for your graphic.

After the image is processed, you can adjust the main contour lines for the eyebrows, eyes, nose, mouth, and face mask (see Figure 5-6). CrazyTalk uses the position of these points to "morph" the picture to match your speech in Skype.

You can test your avatar by having it speak. Click the triangular Play button (see the bottom left of Figure 5-6) to start a prerecorded audio announcement. There are also pause, stop, and volume controls.

As you hear the message played, the avatar's mouth, eyes, and face move in sync with the audio message.

You can tweak the position of the spindle points and contour lines to get a fairly convincing animated avatar.

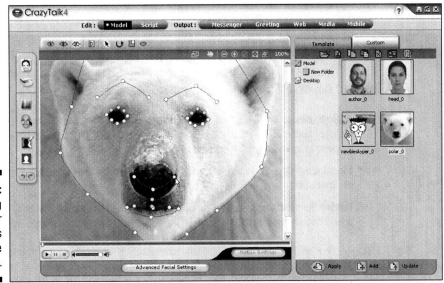

Figure 5-6: Positioning the contour elements and shape masks.

Click the Advanced Facial Settings button to access features that can help you deal with a variety of challenges. Say, for example, that you wanted to make the Mona Lisa into an avatar. Without some modification, you might find that making her image "speak" replaced that indelible smile with — gasp! — a need for dentures! Fortunately, CrazyTalk comes equipped with a set of virtual dentures (see Figure 5-7) and even a collection of eyes.

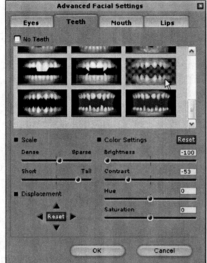

Figure 5-7:
Smile! You
can choose
from a
collection
of virtual
dentures of
carnivores,
pirates, and
normal
humans.

Some of the CrazyTalk facial settings include:

- **Adjustments to the eyes, teeth, and mouth:** You can adjust the lighting levels inside the mouth, or even adjust ends of the lips to give the appearance of a smile or a frown.

- **Head motion:** You can set the head motion to match the energy of your character.

- **Facial formatting:** A nice finishing touch of this software is that you can have the mouth automatically close when the avatar stops speaking.

CrazyTalk Messenger makes it easy to create, test, and customize your avatar. After the avatar is created with Messenger, you can run it in Skype using another software product, CrazyTalk for Skype Video Lite, which is available as a free download (go to `https://share.skype.com/directory/crazytalk_for_skype_video_lite/view` and click the Download Now link).

The Video Lite program connects your CrazyTalk avatar to Skype. CrazyTalk uses the video channel in Skype to broadcast your avatar as though you had a webcam.

To enliven your avatar, you can use a behavior panel with special kinds of emotion smileys (see Figure 5-8). Despite the term *smiley*, you can click these images to show your avatar laughing, crying, frowning, and looking surprised, among other expressions.

Look who's talking

One of the great joys of writing this book is that as authors, we got to tinker and experiment with absolutely fascinating stuff. Our experimentation knew no boundaries. One Sunday, my wife (Susan) and I (Loren) were slaving over a couple of chapters. After she finished her chapter, she decided to take a short but well-deserved nap. I, however, was not through. I decided to make an avatar from a photo of myself. I set up Skype on two computers, one in the study and another on my laptop computer, which has a 17-inch screen. I connected the two systems over Skype and made the avatar of myself show at full screen on the laptop. I placed the computer two or three feet away from where my wife was napping and went back to the study to project my voice into the microphone, calling "Susan! Susan!" She awoke from her nap to a life-sized talking avatar of myself. I said, "Honey, they've turned me into an avatar!" She was not expecting to be awakened by a computer-generated version of me, talking in real time and displaying the animated behavior of smileys. It actually took her a beat or two to realize she was looking at an avatar. Maybe the cartoon hearts floating up out of my image gave it away!

Figure 5-8: This polar bear avatar exhibits a specific behavior as you click the emotion buttons.

During the course of a Skype conversation, you can nimbly switch from one kind of avatar to another by choosing the My Avatars or Default Avatars tab and clicking the image of the avatar you want shown.

Skype Is a Communication Kaleidoscope

Aside from choosing a phone color or style, making a ring louder or softer, or changing a message, personalizing electronic communications has been rather limited until recently. Skype conversations, however, can take many forms at the same time: voice, image, text, and video. Within each of these forms, you have many opportunities to make a unique statement and project your personal style. If you are an artist, you might choose an avatar for yourself as an abstract painting. If you are a businessman, your image can be your company logo. Your choice of personalization can associate you with what you think is important, and you can change your image to reflect a mood, point of view, or occupation whenever you want. The best part is that the choice is yours.

Beyond actively projecting an image or avatar, you can also personalize your Skype environment in the background by adding, or omitting, information in your profile. If you are in business and want a high-profile presence on Skype, the more you put in a profile, the better. If you are the quiet type, you can skimp on the information you put in without taking yourself completely out of the search loop. Personalizing your mode of communication is more than decorating; it's a powerful way to tell the world who you are.

Chapter 6

The Mad Chatter

Chatting with Skype is like flying first class. Skype lets you organize, customize, save, search, and add links to chats. You can invite chatters who aren't online so that they can be free to join the chat circle later. You can send information, files, contacts, and email links in a Skype chat. You can even have private conversations on the sly while you are chatting with a group. The Skype chat landscape is unique, and this chapter is your guidebook.

Set 'er Up and Let 'er Rip

Starting a chat is easy. Pick someone you want to chat with (someone who is online), and follow these steps:

1. **Select a Skype contact from your Contacts list.**

2. **Click the blue Chat button in the toolbar (see Figure 6-1).**

   If you don't see the button, open the View menu to verify that a check mark appears next to the View Toolbar option.

   After you click the Chat button, you're ready to chat! When your chat opens, it's ready for you to chat with the Skype contact you selected.

A confident chat organizer can invite more than one contact simultaneously. To do so, select several (or all) of your contacts in the main Skype window by pressing and holding the Shift key and clicking their names; then, click the Chat button.

Figure 6-1:
The Chat
button in all
its glory.

3. **To chat with your contact, enter your text in the text box at the bottom of the chat window and press Enter to post the message in the chat room.**

A shaded separator bar that contains your name as well as the date and the time of your message appears in the chat window. The separator bars are color-coded so that you can easily spot your messages in the chat, which makes finding your text in the message window easier.

Given the sometimes chaotic, free-for-all nature of chat rooms, you can easily forget your own conversational thread — especially when you're chatting with more than one person at a time. Picking through the messages quickly to find your notes helps you remember the point you are making.

To invite others to join your chat, simply drag and drop a contact from your main window into your message window (see Figure 6-2).

Buddies in a chat room are free to invite their own friends into the chat by dragging and dropping contacts from their own lists. When a new person is added, an alert appears in each chat participant's window. This alert, visible in Figure 6-2, also identifies the user who invited the new participant as well as the time he or she was added. No secrets here!

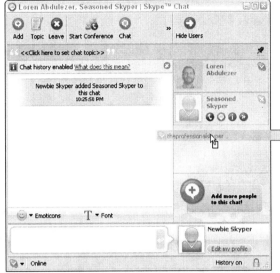

Figure 6-2:
Adding a participant to a chat by dragging and dropping from the Contacts list.

Choosing a slew of contacts for a group chat can be useful. Maybe you want to be sneaky. Maybe you are planning a surprise party for your best friend, and you want to use Skype to make sure that all the people you plan to invite are in on the chat from the beginning so that you don't spoil the surprise. In one chat, you can pick a time, place, gift, entertainment, and any other tidbits crucial to a good surprise while avoiding telephone and email tag. Then you can add your best friend to the chat, pretend that everyone is planning a group run in the park, and have the group enter the date in their calendars. Because your soon-to-be-surprised friend was added midway through the chat, he or she can't see the prior conversation. The surprise is intact and your friend none the wiser. Isn't that what best friends are for?

Why would you have 100 skypers chatting at one time?

As unbelievable as it sounds, you can invite up to 100 other friends or colleagues to a Skype chat. That's a lot of people! Why so many people? Well, here are some things as many as 100 skypers may be gathered for:

✔ Predicting election results from their own congressional districts

✔ Celebrating the exact time the New Year starts in their own countries and time zones

✔ Keeping their own minutes at a Board Meeting

✔ Planning a reunion

✔ Performing in a poetry slam

Mastering the Chaos

Chatting with a friend is easy. When you have chats with three, four, or more people, the conversations quickly become lively, fun, and fast paced. Sometimes the words are whizzing by more quickly than you can type, and it feels more like a video game than a conversation.

In a chat with three or more people, chatters talk past each other all the time. Questions are asked, and before they are fully answered, more comments, inquiries, and observations pop up in the chat window. You wonder, Who is asking, what am I answering, how do we control this conversation? As with a new class of school kids, it's hard to stop everyone from talking simultaneously.

To help manage the chaos, consider designating a chat abbreviation (a word or emoticon) to indicate the end of each complete thought. As a nod to the past, we use the old CB radio ten codes (see the sidebar "The 'ten' codes").

If you are about to launch into a long story, don't wait until the end to post what you've written in the message window. If you press Enter after each line, your chat buddies can read your story as you type it, and maybe they'll be less inclined to interrupt. Just be warned that if you use too many breaks, you may confuse your chat buddies.

The "ten" codes

The "ten" codes are numeric abbreviations for simple sentences developed for use in the military, police, and emergency communications over specific radio frequencies. Citizen's Band (CB) radio users, or hams, adopted the use of ten codes to help understand other CB hams. Problems such as too much static, voices cutting in and cutting out, and difficulty in identifying the speaker and when he or she was done talking resulted in the widespread use of the ten codes.

Bringing order to a chat may lead to using some common signals. Why not the "ten" codes? Useful codes include:

10-3 — Stop transmitting

10-4 — Message Received; Affirmative

10-6 — Busy

10-9 — Repeat your message

10-12 — Stop

10-16 — Reply to message

10-18 — Urgent

Setting a time to chat

Anyone who chats across time zones for business communication knows that coordinating international time zones in a group meeting can be tricky. How do you decide when to meet if one member is in New York, another is in India, and another is in Australia? One way is to agree to set your Skype meeting time to one time zone so that there is never any confusion about when to meet if what you are doing is internationally based.

To set your time zone in Skype, choose File➪Edit My Profile. Select the Show My Time check box and choose a time from the drop-down menu (see Figure 6-3). Of course, if you agree on a time zone that is different from your local zone, you'll have to do a little calculating to figure out how to set your watch, but Skype will be right on time, and you will have eliminated miscommunication and, possibly, missed meetings. It's worth the effort.

Figure 6-3:
Setting your
time zone.

Chatting outside the box

Skype chats are not confined to one message window. A Skype messaging environment is more like a virtual mixer. You can speak to all the members of a group, pair off, have one-to-one conversations, form a new group, swap files with everyone, or share contact lists with some. You can exchange multiple messages by opening as many chats as your desktop, and your concentration, can handle.

Whisper chats

Whisper chats are one-on-one offshoots of group chats. If you are in a group chat and want to whisper to another participant, you can easily open a new, private chat window. Simply follow these steps:

1. **Select a user's name for your whisper chat.**

   You can find a chat partner in the Chat Drawer, a window pane that lists all the users. If you don't see other users, click the Show Users icon in your Skype Toolbar (Windows) or select Window➪Drawer from the Skype menu (Mac). You can have a whisper chat with anyone in your list, even if that person is not on your Contacts list.

Several icons appear in the window of the person you want to contact.

2. Click the Chat button.

A new message window opens in addition to the group chat.

3. Type your message, as shown in Figure 6-4.

Be sure to separate the windows or make them different sizes. If you type a message in the wrong window, well, we're sure you can fill in the consequences.

Blocking a chatter from your Contacts list

Being invited to a chat can be like going on a blind date. Although you may know who invited you, you may not know the other personalities you are asked to socialize with in the chat window. And as with the majority of blind dates (we speak from experience), you may choose not to give out your phone number — or Skype Name, in this case.

If you find a chatter who is objectionable, you can't banish that person from the chat but you can block him or her from your Contacts list so that he or she can no longer have one-on-one chats with you. You can do this immediately in the Contacts window shown during your chat. Simply select the user's name in the Chat Drawer, right-click to bring up the pop-up menu, and click Block This User.

When you block someone, Skype does your dirty work for you and turns away the unwanted skyper.

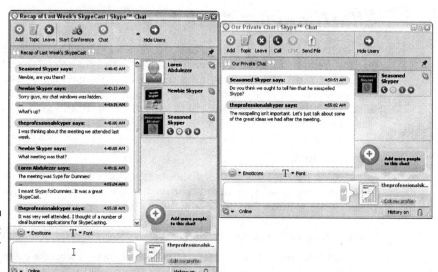

Figure 6-4:
A whisper
chat.

Passing notes around the virtual room

The Skype chat window serves as a virtual conference. As in a face-to-face meeting, you can place a document in front of every participant of a Skype conference — just to make sure that everyone has all the facts straight.

Skype permits the transfer of files instantly to all or only selected participants. If your chat window is too narrow to fit all the toolbar buttons, you can click the double arrow to reveal hidden toolbar buttons and then click the Send File to All button (see Figure 6-5). You now have access to your directory and can open a file and deliver it. What your chatters see is an alert that lets them know you are sending a file.

Figure 6-5: Clicking the double arrow (>>) to see the Send File to All button.

An easy way to transfer files to members of your chat is to drag the file from a folder in Windows Explorer, Windows desktop, or My Computer directly into the chat window. You can even drag an image that's visible from your Web browser into the chat window.

Chat members have to agree to receive your file (see Figure 6-6). If you pick a file to send and realize it is the wrong one, you can cancel the delivery. However, you have to do this quickly, before someone accepts the package.

You can send files only to contacts who have authorized you (they agreed to exchange information upon first contact). So you may be in a chat with five people, but because you have formally authorized only two contacts, you can transmit and receive files only from those two contacts.

Another strategy for sending information in a chat environment is to paste a Web address or an email address in the message window. These are live links that chatters can click to access Web sites. Everyone can check out the same Web page or email the same organization as the virtual conference takes place. To add to the flexibility of distributing information, you can also copy and paste prepared text into a message window, which is even faster than file transfers if you don't care about the document format.

Figure 6-6:
Accepting a
file transfer
from a
Skype chat.

Not every file is meant for all eyes. Fortunately, Skype enables you to send a
file to one person at a time by choosing the double arrow that appears when
you move the mouse into the user's identity box (see Figure 6-7). Clicking the
double arrow opens a vertical menu that includes Send File. This menu enables
a person-to-person file transfer. Alternatively, you can send a file to a single participant in a chat by dragging a file from your desktop or Windows Explorer
directly onto the contact name in the Chat Drawer of your chat window.

Figure 6-7:
Transferring
a file to
a single
member
of a chat.

Bordon Library 2
RENEW ONLINE at www.hants.gov.uk/library or
phone 0300 555 1387

LOVE YOUR LIBRARY

Items borrowed today

Title: Skype for dummies
ID: C0156966528

Due: 20 January 2022

Total items: 1

Account balance: £1.00
23/12/2021 11:10
Items borrowed: 1
Overdue items: 0
Reservations : 0
Reservations for collection 0

Download the Spydus Mobile App to control your
loans and reservations from your smartphone.

A Room with a point of view

You can assign a topic to a chat. Just click the portion of the chat toolbar with "pick a topic" in quotation marks. In addition to a title, you can add emoticons, Web links, and callto:// links (email links) to a topic. To add Web links and email addresses, type them into the topic text box, as shown in the following figure. Because these links are clickable and updateable, chat participants can connect to a Web page containing the chat guidelines, important documents for downloading, or up-to-date information and alerts. To add an emoticon, type its name or copy and paste it from your chat window.

Swapping contacts

File transfer is one of several actions you can carry out on an individual basis. You can also send a list of contacts to other users. Simply choose the Send Contacts menu item and then select the contacts you want to send. Of course, your contacts must authorize your chat partner before he or she can see their online activity, but this process is much more efficient than flipping through the old Rolodex for phone numbers.

Another useful feature of a Skype chat is having each participant's profile on hand. There may be members of the group you have never met. Their profile can contain useful information such as where they live, a Web site for their business, and a phone contact. The old problem of trying to remember everyone's name and what he or she does for a living is greatly alleviated by having access to a profile to review while you're on the chat.

Transferring files and swapping contacts holds enormous potential for efficient workflow. In writing this book, we took advantage of these capabilities, even when we were sitting in the same room. Sometimes the files popped up faster on our screen than we could have handed them to each other across a table. Large files took longer, but after each file was sent and received, we were all on the same page — literally.

Chatting Strategically

Do you ever get a sensitive telephone call at work and find yourself whispering your way through the conversation? Or perhaps you are calling someone on your cell phone and the signal stops. Well, the Skype environment gives you both the opportunity to communicate privately and the choice to type or talk. You can receive a Skype voice call and respond in a chat window. The half-voice, half-chat conversation is a great solution to the open office or portable cubicle work area.

Another reason for a half microphone/half chat conversation is that sometimes you may not have a microphone available. Your sound preferences may not be working properly, so you can't hear the caller's voice (or your caller may be your 86-year-old dad who has no idea what a sound preference is or how he would change it). Having an alternative way to respond to a call provides a simple way to overcome the personal and technical glitches that trip us up on a daily basis.

Leaving a chat

When you want to close a chat, just click the round red button with the white *X* inside in your windows heading bar. The chat window closes but the chat is still active. To reopen the chat window, select the name of the contact you were chatting with and click the Chat button. The same chat window with the same text appears. You can pick up where you left off before you closed the chat window.

Returning to former chats

Just because you left a chat doesn't mean that the chat has disappeared. You can log off of Skype, log back on, and continue a chat that was started previously. To find your former chats, choose Tools⇨Recent Chats and pick the chat you want. Chats are identified by Skype Name and the first line of the chat, or by the topic name if one was chosen and entered. Click the topic and the chat reappears.

If your chat partner is offline when you resume the chat, you may see a gray icon with an *X* inside it in the chat participant list. When your chat partner is logged on to Skype, the icon in the participant list appears green. A white check mark or a little clock symbol appears next to the green icon, depending on whether the participant is available or away. You can still send a message, or receive a chat message, whether you indicate through your status icon that you are not to be disturbed, away, or otherwise unavailable. However, Skype — again doing your dirty work for you — explains why you don't answer.

Bookmarking a chat

Some chats are worth setting aside in a special place for quick access. Skype has a bookmarking feature for these chats. You can bookmark only chats that have a topic, Web site, emoticon, or email in the topic bar; otherwise, the pushpin-style icon used to mark the chat is not shown.

To set up a bookmark, click the pushpin icon in the chat topic bar. When this icon is "pushed in," the chat appears in a special list that you open by choosing Tools➪Bookmarked Chats.

Getting your attention

With Skype, you can choose how you are notified when someone invites you to a chat. You can have a chat window open immediately when the chat starts, or you can have a pop-up alert tell you that someone is waiting for your attention. You can even decide not to have any obvious alert, although you may still like to have the option of knowing that a chat is about to start. Select from the following types of options:

- ✔ **Show alert pop-up:** Choose Tool➪Options➪Notifications and select Display Notification, If Someone Starts a Chat with Me.

- ✔ **Show chat window:** Choose Tools➪Options➪Advanced and select Pop Up a Chat Window When Someone Starts a Chat with Me. Click the Chat alert to expand the window and type away in the text box.

If you don't respond to an alert, your Skype window indicates that one new event has occurred (see Figure 6-8). If you click the chat event link, your chat window opens for business. You may hear a sound alert if you have chosen Tools➪Options➪Sounds and selected the check box next to Play Sounds.

These options (well, two of them) are a little like call screening. You want to know who is calling, but you want to choose your chats carefully. A highly social person with a long contact list can easily be overwhelmed with willing, needy, or just plain gregarious skypers. It's nice to have some control over your time and privacy.

Figure 6-8:
Skype event
alert flag.

Modifying Your Chat Window Dressing

Chat windows contain a lot of information. The toolbar, writing box, topic bar, user list (filled with images, words, and icons), and emoticon and font choice bars are all visible. You can modify your chatscape to make it easier to see and use, and to organize your information.

Hiding your contacts

You may want to simplify the chatscape so that you can see more of the text window. Click the Hide Users button to eliminate the Contacts list and expand the user text window. You still know who is chatting, and you can bring your Contacts list back at any time by clicking the Show Users button.

Changing your text size and style

Changing the font and increasing the font size may be both decorative and practical. Large fonts are easier to see. Fonts with serifs are more formal as well as easier to read, but they take up more space. You can change the fonts by selecting the Font pop-up menu just above the writing box in your chat window (see Figure 6-9). You can experiment with each font to see which is best for you. Note that changing fonts may replace the shaded separator bars with outlined placeholders (which may or may not be empty), but forfeiting shaded separator bars is a small price to pay for legibility.

Any font changes you make in your chat window are for your eyes only. Other chatters do not see your font selections.

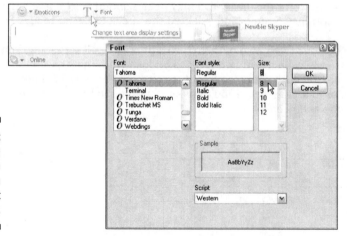

Figure 6-9:
Adjusting
font type
and size in
your chat
window.

If you change your font size and style and are totally miserable, don't despair; you can restore the default font. It's Tahoma, 8 points, Western script.

An abbreviated conversation

Chats are so fast and furious that people have created their own chat shorthand. The following dialogue (see the figure) among three skypers quite literally takes this conversational grammar to an extreme because it is a conversation made up entirely of abbreviations and acronyms.

People frequently mix in chat abbreviations in their conversations. When the same circle of friends converse all the time, there is a tendency to spontaneously evolve their own private shorthand.

To see the conversation in full English, flip the book upside down.

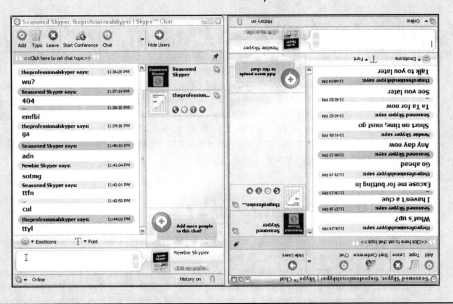

Opening your windows

Skype sets the area you use to write text as the smallest box in the chat window. It doesn't have to stay that way, though. You can "grab" the top of the text window and pull it up to have more space to write and see what you have written. Enlarging the space allows you to view paragraphs of text you are editing and are about to send to your fellow skypers, instead of being able to view just a few lines in the text input area.

As with other Instant Message programs, you can expand the chat window to make it a full screen.

Another way to clean your window is to right-click in the body of the message window and select Clear Messages, which wipes out all the messages in your window. Fortunately, your neatness doesn't clean out anyone else's windows.

If you really want to pare down your messages to their simplest form, change the chat style in the message window to IRC (Internet relay chat). Choose Tools⇨Options⇨Advanced and select IRC-like Style from the Chat Style to Use menu. Deselect the Show Timestamp with Chat Messages option. Your chat window will hold more messages in a smaller space (but might be harder to read).

Chatting in the Past, Present, and Future

> Chats persist. That is, they don't necessarily disappear when you leave them; they come back when you call them; and they can go on and on. In contrast to other instant messaging systems, Skype chats are like social diaries. They record and remember everything you wrote and everything written to you unless you make an effort to eliminate them by clearing windows or limiting how long they are saved in history. Read on for more about saving and deleting chats.

The never-ending chat

A Skype chat can last forever. A group of chatters, passionate about a particular topic, can form a chat club, forum, or special interest group. If they bookmark the chat, then each time someone writes a new comment all participants are instantly notified in their chat taskbar, even if all the chatters have closed their chat windows. In fact, when a new message is added, all the bookmarked chats "wake up." To make a chat an ongoing chat, each member of the club must leave his or her Skype software running. The best part is those 3:00 a.m. epiphanies when all the points you wished you'd made in the middle of the chat come to you. Now you have an audience.

When the chat is over. . .

All chats are archived. It is a default setting in Skype to save every last bit of chat history. You are not forced to do this, and turning off the chat history is a privacy feature (to the extent that chatting is a private act). But the idea that everything you write, and everything everyone writes to you, can be saved, published to an RSS feed, added to a blog, or just printed, outrivals the most ambitious diary ever written.

Preserving your chats "forever" is easy. Choose Tools⇨Options⇨Privacy and select Forever from the pop-up menu that appears. Of course, you can also choose No History, 2 weeks, 1 month, or 3 months, but there's something to be said for having the ability to save your words forever (or at least until you get a new computer). In the same Privacy window, you can click Clear Chat History and wipe out your entire archive. The catch is, your chats probably exist on someone else's computer, and maybe that person is an avid chat collector. Chats are saved locally, not on any central server. All chat participants have equal-opportunity archiving capability.

Searching, saving, and sending past chats

You are just finishing a chat that lasted for hours — a discussion among three international offices about the state of their factory construction. Now you have to take all the information and organize it for a presentation. Who said what, when, in what order, and how do you find it?

Searching your chats

In a way, Skype conference chats are efficient in that they save all the information in the exact language each person used. All the statements, re-statements, corrections, corroborations, statistics, Web links, and even emoticons are frozen in time. You can relive the whole experience as many times as you want (which we call an *added benefit,* for the purposes of this book).

To collect bits and pieces from this massive conversation, right-click in the message window and click Find to open the Find dialog box. You can search by word, time, or Skype Name. Because the chat may go on for pages and pages, the Find utility lets you search up or down (from the end or the beginning of the chat).

To find a specific chat, select the Chat History tab and scroll down to pick a topic or first sentence, date, or contact; then, double-click to reopen the chat.

Saving your chats

Transform your chats from words in a chat window to a document or presentation by selecting all the messages, copying them, and then pasting them into a text editor. These documents retain the messages, timestamps, and author of each statement. Some skypers have turned their chats into RSS feeds and podcasts, passed them through text-to-speech programs to make them audible, or even used them as captions for a movie illustrating the chat topic.

Transcripts of chat sessions are on your hard drive. If you switch machines, such as between home and office, the session is associated only with the machine you were using. If you need to have the transcripts, Skype contacts, and in fact all your files centralized in one place, you can run Skype "on a stick" (see Chapter 9 for more information on USB Smart Drives, which are thumb-sized drives that have enough capacity to contain the Skype program and all your personal Skype information).

Sending your chats

Message texts can be pasted into email documents, word processors, presentation tools, new Skype chat windows, or Web authoring tools. When the chat is preserved in its new form, it can be emailed, posted on the Web, entered into a blog, or sent in any way digital documents are shared. If you want, you can print them and send them by snail mail.

The three rules of "chattiquette"

✔ Rule 1: When your Skype buddy is on the other side of the world and you want to call at an odd hour, use the chat window to notify, send information, make an appointment, or otherwise alert him or her because you know a call would be disruptive.

✔ Rule 2: Use a text editor, such as Notepad, to compose long paragraphs and then copy and paste the text all at one time. This way,

you avoid the carriage return breaks in a chat window and make the chat more efficient because you won't make your chat buddies wait as you type.

✔ Rule 3: Use chat abbreviations only when you know they won't baffle your chat buddies. Use them liberally when everyone is on the same chat wagon.

Chapter 7

Skyping Eye to Eye: Skype with Video

*I*n the opening scene of the movie *2001: A Space Odyssey* (produced in 1968), Dave Bowman is talking with his wife and children over a video phone. As the movie title indicates, the story takes place in 2001. Back in 1968, we waited with bated breath for the advent of video phones. We waited a long time: The year 2001 came and went with hardly a trace of video phones becoming a reality. Finally, history is about to catch up with science fiction, thanks to Skype.

This chapter walks you through choosing the right equipment and environment for setting up Skype with video. We also take you through a webcam selection, connection, detection, and even a section about video protection. Okay, we know when to stop, or, to quote the sportscaster Warner Wolf, "Let's go to the videotape" (in this case, let's get started with Skype video).

Enhancing Your Conversations with Live Video

The addition of video to a call may make the difference between touching base and enjoying a true visit, describing an item for sale and clinching the deal, or worrying whether someone is doing okay and seeing for yourself. Coming together in agreement, truly communicating intentions, and clearly describing objects are all circumstances that benefit from the ability to see eye to eye, literally. Videoconferencing can have some unexpected bonuses:

- ✔ **Webcams enable group participation:** By setting up a webcam, a family can join in on a Skype conversation without having to pass headsets around from person to person.

- ✔ **Webcams eliminate wired microphones:** Good webcams have built-in microphones, so you don't need to use the computer's internal microphone or add an external microphone.

- ✔ **Webcams may eliminate echoes:** Webcams with noise cancellation features eliminate the voice echo your Skype partner may hear if you don't have a headset. This noise cancellation feature lets you speak, unfettered, while using Skype video.

- ✔ **Webcams eliminate wired headsets:** A surprising advantage of connecting a webcam to communicate is that a good webcam takes the place of tethered headsets. Sound is routed through the webcam speakers instead.

If you don't have a webcam, you can still receive a video call. Skype software lets you see video transmitted from another user even if you don't reciprocate.

Considering Types of Webcams

Before you use Skype video, you have to beg, borrow, or buy a webcam. Picking a webcam can make the difference between a great videoconference and a frustrating one. First, determine how, where, and why you are using videoconferencing. Then pick a unit that best matches those needs. If you make video calls at a desktop in your office, you can get a unit that is larger and can be moved around, placed on a shelf, or mounted on an office wall. If you are on the move and need to be totally portable, choosing a small unit that attaches to your laptop is a better choice.

Newer computer models are beginning to include built-in webcams, such as the Apple iMacs and MacBook Pro laptops. A built-in webcam is the most compact way to include video. But if you don't have that option, you'll have to do a little research and make a few good choices. The following section should help you with your webcam buying spree.

Standalone webcams

The standalone webcam rests on a platform or a boom (a small pole that supports the webcam), or it nests in a stand. Standalone cams are usually larger and some may have more features than those smaller units that clip onto a laptop or flat screen, although high-end clip-ons are almost indistinguishable in performance from high-end standalones. If you have a dedicated video-conferencing area, have multiple people at your conferencing station, or want to include a fair amount of items in your video viewing screen, then standalone webcams are for you.

Logitech's Quickcam Orbit MP (see Figure 7-1) is an example of a feature-rich standalone webcam. The camera is a shiny orb with a wide-angle lens that perches on a variable-height post or small stand. Its ultra-wide-angle lens takes in a large area around you. You may have the corner of a room, or a curved conference display set filled with your latest widgets on floor-to-ceiling shelves. The camera can include your widgets along with you and your sales partner in the mix. This webcam also follows you around your work area using a capability called Face Tracking. The Face Tracking webcam has some cool advantages:

✔ **Face Tracking allows the camera to fix its lens on the dominant face in the picture (or the most active one).**

✔ **With Face Tracking, as you move, so will the camera:** Logitech's Quickcam Orbit MP webcam actually has a motorized lens that physically moves as it tracks you.

✔ **Face Tracking accommodates both pan and tilt motion as you move across and up and down:** Some webcams move within a 189-degree field of view and a 109-degree tilt up and down.

✔ **Face Tracking adjusts the webcam focus as you move closer or farther away:** If you move closer, the image is tight, framing your face. If you move back, the webcam adjusts its zoom to reveal more of the room. As you move, the webcam changes focus dynamically. Having a camera that makes its own zoom and pan decisions is, well, actually a little spooky!

See the section "Follow this face, or that face, or those faces!" later in this chapter for more about Face Tracking.

Clip-on webcams

The clip-on webcam perches on the top of a flat screen and is perfect for a portable videoconferencing setup. One of the big advantages of a clip-on webcam is that it overcomes the common drawback of the casual videoconferencer. That is, most people stare at their screen, at the video image of the person they are talking with, rather than look into the camera. Unfortunately, most cameras are placed in an awkward position, so it doesn't make sense to look into the lens. It makes much more sense to watch the person on the screen. But then you have two people not quite making eye contact in a medium that's supposed to connect people face to face!

Although anchoring a clip-on at the top of a screen helps aim the lens, some clip-on webcams pivot to allow some subtle adjustment of the lens. You pivot and aim the webcam at your face while you test the software settings for your clip-on. After it's set, you don't have to worry about the awkward stare into the computer.

Figure 7-1:
The
Logitech
Orbit MP.

Some clip-on webcams, such as the Logitech QuickCam for Notebooks Pro
(see Figure 7-2), have a digital Face Tracking feature (rather than a motorized
lens tracker, as described previously) that keeps you in focus as you move
toward or away from the lens. The camera doesn't move, but the software

detects faces and makes adjustments to keep them front and center. If you are talking and suddenly lean back in your chair, the image zooms out to show you and the room around you. When you lean forward, the software tries to keep you in the viewing area by zooming in and eliminating the background.

Figure 7-2:
The
Logitech
QuickCam
for
Notebooks
Pro.

Photo courtesy of Logitech

If you don't want to use a wired headset and microphone, you can take advantage of your webcam's microphone and noise cancellation feature without having the webcam active (heaven forbid you have a bad hair day). In this mode, the webcam's microphone can continue to work even if the webcam settings in Skype are not enabled.

Specialty webcams

The most common use for a webcam is to see the person you are calling. But there are other webcams that have more specialized missions. Our favorite is the ProScope USB microscope (see Figure 7-3), available for both PC and Mac platforms. After you download and install the proper drivers that come with this device, it works just like a webcam, with a little twist. The microscope can, of course, take extreme close-ups of objects and materials. Some microscopes come with a variety of lenses to increase the power of the "macro" photographic image. There is even a high-res version of the scope (find out more at www.proscopehr.com/index.html).

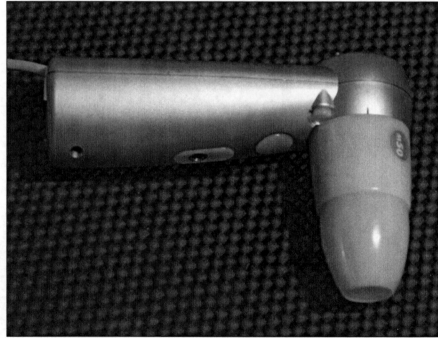

Figure 7-3:
The
ProScope
USB
microscope.

Photo courtesy of Loren Abdulezer

Why would you want to use a USB microscope? Perhaps you are a coin collector (see Figure 7-4) and would like to make some trades, purchases, or sales within your Skype coin club. Connect the microscope, turn on videoconferencing, and put your wares on display. The same can be done for other collections; obsidian arrowheads, cameo jewelry, scrimshaw art, and postage stamps are just a sample of collectibles you can show in detail over Skype using a USB microscope.

Figure 7-4:
A USB microscope can transmit close-up images of coins and jewelry, for example, over Skype.

Another use for a USB microscope is the transmission of live video streaming in field research. If you are a biology teacher, equip your students with a laptop, a USB microscope, and an assignment to capture images of moss samples from a local park. The USB microscope has a built-in light source that will illuminate a bit of moss on a rock or tree bark. Students in other schools anywhere in the world can receive the video of local flora and add it to a database of moss samples for research projects.

Transmitting live video from the field is not only inspiring for students and teachers alike but also arms the students of today with tools for thinking about how to design the laboratories of tomorrow. Skype videoconferencing combined with USB microscopy makes it possible to routinely use field microscopes in disciplines such as forensics, and teams of people spread over thousands of miles can collaboratively engage in problem solving in real time.

"MicroSkype"

There are many ways to use Skype and a USB microscope. Here are some kinds of activities you may want to try.

- Show close-up images of the mint mark of coins for sale.

- Video the undersides of house plant leaves so that the local botanical garden can identify the mites eating your precious rhododendron.

- Show the engraved image on a collectible stamp.

- Focus on the artist's signature on the bottom of a porcelain figurine.

- Transmit the carat marking on a piece of gold jewelry you are showcasing.

Understanding Webcam Features

Image size, automatic and manual pan and tilt control, and autofocus are all important features of a good webcam. To help you decide which webcam to use, here are some things to consider.

Sharpening those pixels

The images rendered by most inexpensive webcams have a classic security camera look — jumpy black-and-white images that have more in common with Charlie Chaplin movies than Paramount Pictures. However, the newest webcams have made tremendous improvements in transmission, color rendering, and image resolution. Some give you a choice of image quality or size. If you pick one with a better quality image, though, you may run into some transmission problems. Better pictures contain more information that must be pushed through those Internet cables to reach your Skype buddy. The general rule is to pick a size and resolution that's not too large and not too small, but just right. Experiment with your equipment to figure out which size works best. If transmission is choppy, reduce the picture size in your webcam software preferences. To change your video image size, open the software installed with your specific webcam. You cannot change the image size within Skype software. It is important to explore your webcam software before you make a video call on Skype so that you can be prepared to troubleshoot if there is a problem.

If you are on a Skype video call and you don't see yourself in your video window, reduce the pixel resolution of your webcam software in the video settings preferences. Chances are, you'll reappear!

Follow this face, or that face, or those faces!

Motorized and "intelligent" Face Tracking is a feature that frees you from end-lessly adjusting your webcam to keep your face in the viewfinder. The digital or motorized face tracker enables you choose to track one person or detect several people at a time (see Figure 7-5). When a group is in the picture, the motorized face tracker adjusts to try to try to keep everyone in the frame, even if everyone is moving in different directions. Logitech's Quickcam for Notebooks Pro software lets you choose Face Tracking for a single user or multiple users to follow an individual or a group.

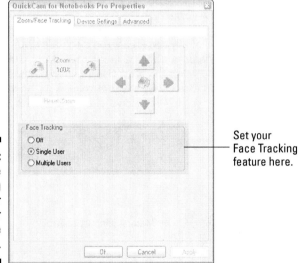

Figure 7-5: Setting Face Tracking mode for single or multiple users.

Set your Face Tracking feature here.

Face Tracking doesn't just make webcams easier to use with Skype. Software that tracks facial features has the potential to perform other kinds of func-tions as well; for example, you may be able to have the software recognize a face, remember a face from a former videoconference, or look up and match a face from a library or photo album. Facial recognition is a handy feature for those of you who never forget a face but never remember a name. You may also be able to analyze the facial expressions and head motions of individuals in different cultures (great for businessmen and diplomats), follow an individ-ual's eye movements, or even have the software read lips. All these high-tech features are in the works and will eventually add power to our Skype video communication.

Toggling into focus

If you are a little hesitant to let your camera follow you around, you can always activate the manual pan and zoom option. To do so, follow these steps:

1. **From the Skype menu, choose Tools⇨Options⇨Video⇨Test Webcam⇨ Webcam Settings.**

2. **Select the Off radio button to stop Face Tracking.**

Click the arrows in the Webcam Settings window that appear above the Face Tracking radio button choices to move the camera lens in any of four directions as well as to zoom in and zoom out (see Figure 7-6). Disabling Face Tracking in the Webcam Setting window is preferable when you want the camera to stay focused on an object that you are describing, or if you want to control who is displayed in the video as you speak. If your mom were sitting beside you in front of a webcam as you spoke to your niece, you would probably want the camera on your mom's reaction when she finds out that a great-grandchild is on the way. Use the manual pan toggle to make sure that the camera is exactly where you want it . . . on your mom's broad smile!

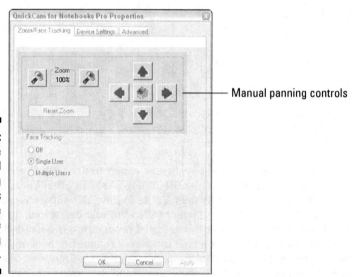

Figure 7-6:
Available
manual
panning
controls
activate
when Face
Tracking
is off.

Manual panning controls

Shopping for Webcams

While you're webcam shopping, it's a good idea to review the list of features published for each webcam you're considering. Compare that list with the following items to get the most flexibility out of videoconferencing with Skype:

- Wide field of view
- Microphone included with camera
- Microphone contains noise cancellation capability
- Minimum lens distortion
- Low light boost
- Manual pan and zoom control
- Face Tracking software
- Avatar creation software included
- USB 2 enabled for fast transmission
- Wide-angle lens

The less you have to fiddle with your webcam, the smoother your video conversation will be. A webcam with a fixed focus nails you to the chair. You can't move out of range, reach for a book, or lean back in your chair without slipping out of camera range. A webcam with a wide field of view that tracks your movement, keeps your features on-screen even in dim light, and doesn't distort your features makes video chats a more natural activity.

Webcams with built-in microphones, especially noise cancellation microphones, give you a definite advantage. Microphone webcams eliminate an extra piece of hardware with extra wires that take up precious USB ports on your computer. If making a video call forces you to plug in a webcam, a microphone, and headphones as well as deal with the tangle of wires, plugs, and devices, you may just decide to forego the whole experience. One good webcam combines all these devices, simplifying the setup and letting you concentrate on what you want to say, not what you need to plug in to say it.

 Often, you can take advantage of your webcam's noise cancellation microphone even if you don't enable video. You can choose your webcam as the sound input device in your computer's sound control panel and not even have to set up a separate microphone! To do so, follow these steps:

1. **Choose Start⇨My Computer⇨Control Panels⇨ Sounds and Audio Devices⇨Audio.**

2. **In the Sound Recording section of the Audio window, click the drop-down list and select your webcam.**

Installing and Setting Up Video for Skype (On Windows)

Now it's time to roll up your sleeves and plug in that camera. You need to install some software and then configure it before you launch your on-screen debut.

Installing your video driver

Your webcam should come with its own installation software. Although each brand is different, all involve some common steps to install the software drivers. Following are some tips to consider.

Turn off your anti-virus and anti-adware software before installing new video drivers. You can turn them back on after the video installation.

The webcam you may be using may not have the most up-to-date video driver for your computer and operating system. Check for updates from the manufacturer on its Web site. (Hint: Look at the support and download pages for updated drivers.)

Don't plug in your webcam before you install the video driver. In almost every webcam installation, you need to wait until the software prompts you to plug in your device (see Figure 7-7).

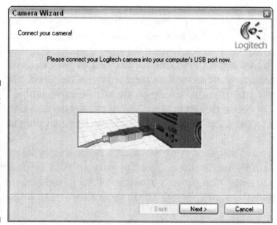

Figure 7-7: Wait until you're prompted *before* you plug in your USB webcam!

If you are encountering technical difficulties with your video, go to Appendix B of this book; you may find some help there. If the installation did not work, simply uninstall the video software and start over with the proper sequence.

Setting Skype video options

To configure your Skype video options, choose Tools⇨Options⇨Video (see the left side of Figure 7-8).

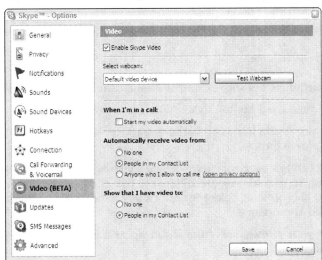

Video options require that you have a video device, either a webcam or software such as CrazyTalk (which animates a character's mouth and mimics your lip movements as you speak). To use video, you have to select Enable Skype Video before continuing. You should deselect the When I'm in a Call: Start My Video Automatically option. You probably should also set the Automatically Receive Video From option to People in My Contact List. Here's a list of Skype options you may want to set:

Category	Option
Privacy	Allow Calls from My Contacts
Video	Enable Skype Video
Video	When I'm in a Call: Automatically Start My Video
Video	Show That I Have Video to People in My Contact List
Advanced	Automatically Answer Incoming Calls

Skype video is very stable. However, Skype is constantly improving its software and therefore continues to label its video feature in Beta mode, as you can see in Figure 7-8.

Checking your sound settings

You'll want to make sure that you're actually piping in sound through your webcam and that the volume level is adequate. Many webcams have visual sound meters that light up as you speak so that you get feedback concerning your volume. One reason to check your webcam's sound is that you may have more than one sound input device attached to your computer. If you're unsure of which one is active, just tap the microphone — you'll find out soon enough!

To adjust your volume:

1. **Choose Start⇨Settings⇨Control Panels⇨Sounds and Audio Devices.**

   The Sounds and Audio Devices window appears.

2. **Click the Audio tab to open the Sound Recording window; there, click the Volume button.**

   In the window that appears, you see a virtual sound mixer with four volume sliders. Among the sliders are two choices for audio input, Line In and Microphone. Select the check box beneath the volume slider labeled Microphone

3. **Using your mouse, move the slider up or down to raise or lower the volume.**

Making a video call on your PC

Now you'd better comb your hair because you're ready (we hope) to become a video star. Glance in the mirror to check your smile, clear your throat, adjust the camera, and take a deep breath; you're about to go live on Skype!

To make a video call on your PC, follow these steps:

1. **Select a contact from your list in the main Skype window.**

2. **Click the green call button.**

3. **When your contact answers, click the little video icon located at the bottom of the Skype call window.**

That's all there is to it! You're on the air in your very own show with your very own fans (even if those are just Mom, your kids, or Dennis from the office).

 It's a good idea to test your webcam before you start your video call. From your Skype menu, choose Tools⇨Options⇨Video. In the Video Options window, click the Test Webcam button. Your video window will appear with a live image. If you don't see anything, make sure that you have selected the Enable Skype Video check box and your cables are all plugged in!

Installing and Setting Up Video For Skype (On the Macintosh)

Some Macintosh computer models, such as the iMacs, have cameras built in, so there is no need to install a separate webcam and special drivers. Macintosh laptops and towers can accept firewire webcams (webcams that connect with a special firewire cable for fast video data transmission), such as the iSight camera, also without any additional software installation. Skype works seamlessly with the Macintosh video drivers to detect iSight as well as built-in cameras.

Not all webcams work with Macintosh or PC computers. Make sure that you read the system requirements on the webcam package before you buy it for your Skype video setup.

Setting up Skype video on the Macintosh is simple. The following steps use iSight as an example, but the steps should be similar for others.

1. **Choose Skype➪Preferences➪Video.**

   The video configuration window opens (see Figure 7-9). If you don't have a built-in camera, plug your iSight webcam into a firewire port and turn it on by opening the lens.

2. **From the Camera drop-down list, select your camera if it is not already detected.**

3. **Select the Enable Skype Video check box.**

4. **Choose your video options by selecting the check boxes or radio buttons (as the case may be) next to each choice, as follows:**

   • **When I Am in a Call:** It's best not to check the option to start your video automatically. Give yourself a chance to compose yourself before appearing on camera.

   • **Automatically Receive Video From:** You can be super cautious and choose no one, or you can allow video only from people on your Contacts list.

   • **Show That I Have Video To:** You can choose to keep your video capability a secret and choose no one, or reveal it only to people on your Contacts list.

5. **Click the red gel radio button on the far-left corner of the window title bar to close your Preferences window.**

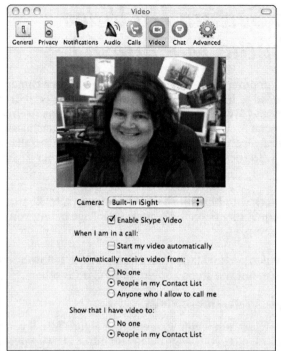

Figure 7-9:
Mac video
setup.

There you have it! You're all set to make a video call using your Macintosh, and you can communicate both with other Macintosh and PC users with this video capability. To make a video call on your Macintosh:

1. **Select a contact from your Contacts list in the main Skype window.**

2. **Click the green call button next to your contact's name.**

3. **When your contact answers, click the little video icon (see Figure 7-10) located at the bottom of the Skype call window.**

   Your video will be transmitted to your contact. When video is live, your video icon turns an aqua blue.

Skype doesn't automatically broadcast your image, even if you enable video in your options. The default setting in Skype is protective: Nobody has to see you in your pajamas when you're making or receiving a Skype call. To make yourself visible, you have to click the blue video icon when a call comes in (see Figure 7-10).

Skype "Webcammandments"

We've all seen the caption rolling across our blank TV screens: "Please pardon the interruption. We are experiencing technical difficulties." Skype video is your show, and a little attention to lights, sound, connections, and camera placement helps your viewers enjoy it more. Use the following "commandments" to help you keep your show on the air:

- ✔ Plug in your camera.

- ✔ Plug in your camera before launching Skype.

- ✔ Plug in your microphone before using Skype (unless your video camera has a built-in microphone).

- ✔ Point the camera the right way.

- ✔ Turn on a light and let it shine on you!

- ✔ Turn up the sound. Make sure that your sound setting isn't on mute.

- ✔ Make sure that your image resolution is not set too high; otherwise, you might see funky pixels and colors. You can change your image resolution within the specific video software installed with your webcam. Not all webcam software offers a choice of resolutions, so you may not have to worry about changing this. But if you have a choice, a video image that is 640 by 480 pixels and millions of colors may be a little rich for transmission bandwidth.

- ✔ If you don't see or hear anything, close Skype, plug in all your devices, and start Skype again.

Figure 7-10:
Skype video transmission doesn't occur unless you click the blue video icon.

Blue video icon

Choosing your video "set"

It's a good idea to pick a place, a "movie set," for your videoconference. Although your hair might not cooperate for your on-screen debut, you can at least get the backdrop right. Here are some ways to get "set" for video:

✔ **Select an all-purpose background:** The simplest set is a solid-colored background. Quiet backdrops such as curtains or undecorated walls actually make transmitting video a little easier because there is less visual information to send. However, with high-speed lines and more efficient webcam software, you can add more interesting backgrounds without degrading the video signal too much.

✔ **Create a mood:** A good backdrop for your video setting might be a library wall. Another is your collection of photos or artwork on display. On the other hand, if you're teaching online and need to demonstrate a lesson — perhaps you are teaching a Sign Language class, for example — then a simple, blank backdrop will prevent distractions for your viewers.

✔ **Lighting your set:** Having a well-lighted area keeps you from being lost in the shadows. Lighting brings out color and detail of objects and individuals. Sensitive webcams can boost low light, but they need to have at least some light to boost.

You can move a lamp stand close to your "broadcast station" if you don't have enough natural light. Arrange your light source to illuminate your face. Fluorescent lights, usually mounted on the ceiling, cast a greenish glow (not very flattering), whereas tungsten lights, such as the ones usually used in table lamps, warm your skin tones.

Although natural light is best, it's available only during the day, of course. If you do have a source of natural light, take advantage of it, because you still need a little more light than whatever your computer screen provides.

During the day, you can set up a laptop and portable webcam near a window, but be careful not to aim the camera at the window unless you want to appear in silhouette. Shift your chair, or move the angle of your laptop or desktop to take advantage of a light source such as a window behind a desk.

✔ **Control background sound:** Another consideration for an effective video set is background sound. If you live near a road or highway, you might want to move away from the window to eliminate all that motor noise. If you are near a construction site, calling in the evening, after workers go home, is a better choice.

✔ **Choose a set outside your home:** With wireless Internet access in public parks, cafes, or even libraries, you can have a video chat from your favorite hangout. Keep the location consistent, though. Every time you pick a new place to set up your webcam, you need to find a wireless connection, adjust your position to take advantage of the light, and accommodate background noise. Showing up at the same place at the same time shortens setup time. If you pick a venue with a lively, noisy background, dedicated headsets are required so that you can hear and be heard.

A picture is worth a thousand words, but you need to hear the words as well as see the picture; otherwise, videoconferencing isn't much fun. Visual "noise" can be as disruptive as any other brand of noise. On the other hand, all that background activity may be just what you want your caller to see. There's no rule that videoconferencing needs to be just face to face. Determining the view your webcam shows makes you a director, of sorts, so be creative as you express yourself. Take advantage of the capability. A few preparations can make a big difference in your experience.

Your videoconferencing environment is a balanced combination of backdrop, light availability, and camera model. Some more expensive webcams are much more sensitive to low light and transmit significantly better at night.

Video in Action

One fun aspect of using Skype video is that it doesn't have to be an isolated tool. Combining Skype video with other Skype features opens even more possibilities.

Video with chat

You might think that having a chat window open during a video session would be redundant. After all, you can see and talk — why would you want to write, too? Here are several great reasons:

✔ **Exchange paragraphs of text:** If you are collaborating on a presentation and you want to extract a paragraph from an existing document, you can copy and paste your work into the chat window as you are talking with your partner; this way, you can show the information.

✔ **Pass a Web link:** You can give someone a Web link by just pasting the URL into the chat. Sometimes there's no substitute for connecting straight into a site. Doing so clarifies the conversation because showing a Web site, image, or paragraph is easier than describing it.

✔ **Assist in conversations with multiple languages:** If you are conferencing with someone who speaks a different language, the chat window can serve as an instant caption to clarify, translate, or define confusing phrases.

You can find out much more about using Skype chat features in Chapter 6.

Video with file transfer

Video lets you see your fellow skyper, so it's natural for both of you to want to see what you are both talking about. Unfortunately, holding up an article, letter, or any printed text in front of a webcam is likely to produce a slightly shaky, slightly fuzzy image that is more than slightly illegible. Instead of holding up a paper, transfer that file (see Chapter 6) if you have it in digital form. If you don't have a digital file, you can scan the article in advance of your video call and then transfer it. You don't have to open your email to send the file, nor does your recipient have to open his or her email to download an attachment containing the file. We've found that transferring a file while we are connected through Skype video does not slow down the transfer rate. Mention it, send it, and show it. It's almost like being in the same room!

Creative Uses of Skype Video

After your get your video gear ready, you can find many ways to use it beyond a cozy video chat with Grandma. Following are a few video scenarios to consider.

Setting up a "NannyCam" or SurveillanceCam

With Skype, you can set up a "NannyCam" to make sure that your kids at home are fast asleep by 8:00 p.m. Here are some quick and easy ways to set up a surveillance camera at home:

- **Create a special Skype Name for your NannyCam:** Be sure to set to set your Skype Sign In settings to Sign Me in When Skype Starts and Start Skype When the Computer Starts. This way, if your computer reboots, your NannyCam starts up automatically.

  If you are already logged in to Skype, you can change these settings by choosing File➪Sign Out from the Skype menu. A login page appears where you can set these options.

- **After you sign in with your NannyCam Skype Name, add your regular Skype Name to your Contacts list:** The Contacts list for the NannyCam consists of only one person, namely, you. Of course, you may want to include your spouse's Skype Name alongside yours. Whatever the case, the list is very short.

- **In your NannyCam's privacy options, be sure to set the Allow Calls From option to Only People from My Contacts:** Setting this option ensures that only you and your spouse can connect to the NannyCam.

✔ **Enable video and set it to start automatically:** In your Skype Video Options window, select Enable Video and then select the check box next to Start My Video Automatically. In the Advanced Options, set the Call option to Automatically Answer Incoming Calls.

Test your NannyCam by connecting to it over Skype from one of the other computers in your home, or even from work.

Be sure to adjust your webcam to point to the specific area of the room you're watching, and make sure that the lighting level is adequate. Remember, most webcams give some visual signal that they are transmitting, such as an LED that lights up. You can adjust the audio volume levels for the computer speakers next to the NannyCam so that you have a little PA system, or you can turn off sound altogether.

This kind of setup can give you some peace of mind when you are traveling and want to make sure that everything at home is safe and sound. Best of all, this is a simple and very affordable home security solution for homeowners. Get yourself an inexpensive webcam and dust off that out-of-date computer taking up space in your closet. You won't have to worry about whether it has enough power; the webcam and Skype are the only applications you need it to run.

Using video in the field

Video with Skype does not always have to entail having two people who are visually connected, one on each end of a live conversation. A USB microscope attached to your portable laptop and Skype on a Wi-Fi network can give you a close-up view of the world. There are plenty of practical uses of Skype and video that aren't limited to headshots of people talking to one another. Read on for a few ideas.

Getting the scoop, and then some

Field reporting using a laptop that connects to the Internet from one of the phone company's PCMCIA cards, equipped with Skype and a small video camera, can be a powerful tool for journalists. Whether a reporter writes for a newspaper, magazine, e-zine, podcast, or blog, the combination of laptops, webcams, and Skype is the equivalent of a TV studio on a shoestring budget. Transmitting live pictures, sound, and text from the middle of an event eliminates any time lapse between what happens in the world and how quickly people hear about it. Television has long been able to bring immediate news to a broadcast medium. Skype video, audio, chat, and file transfer put this same power into the hands of individuals. An enterprising Skype reporter can take his or her live feed and make it into a podcast, or even a "vodcast" (a video podcast) for large-scale distribution. Reporters can even connect to a newsroom with a live Skype video call and have their transmission broadcast on television.

Searching for answers with Skype and a scope

Researchers can use Skype video to deploy a group of biologists (or budding student biologists) to collect data, information, and images in the field. Connect a USB microscope to each field laptop and transmit pictures of soil, water, moss, and lichen samples over Skype video. The research group can coordinate field research efforts to their best advantage and send images to remote laboratories, schools, and colleges.

Any field-based research benefits from video transmission tools. You can be digging for arrowheads, pottery shards, or other archeological artifacts with a live webcam or USB microscope serving as a research tool. You can enable a remote expert on the receiving end of the video stream to perform an instant analysis. The scientific tasks can be distributed among field, lab, and academic sites. More researchers can be involved, with less disruption to their own schedules. The research efforts can even be multinational.

Skype is currently limited to two video partners at a time, but with some third-party software programs (check out Festoon at www.festooninc. com), video conferencing among more than two people at a time is possible, which multiplies the collaboration efforts of any field-based projects.

Tuning into the business of Skype video

Use Skype video to sell your wares, collaborate on developing a product, or inspect a custom-made prototype before shipping it out to the manufacturer. In business relationships, seeing the person you're doing business with is helpful, but seeing a product is crucial. Skype video is a quick way to make business communications more efficient and speed the workflow and production environment.

On a small scale, you can set up a jewelry cam to transmit pictures of items, such as rings and earrings, to potential customers. Use a USB microscope to take a closer look at the engraved markings on various items as you talk about each piece. A static picture is helpful to a buyer, but video is dynamic. You can respond to a request to focus on a stone or a setting, thereby satisfying a buyer's curiosity and overcoming any reluctance on the part of your customer to make a purchase.

Chapter 8

The Ins and Outs of SkypeIn and SkypeOut

* *

In This Chapter

▶ Understanding and using SkypeOut

▶ Buying SkypeOut credit for international calls

▶ Understanding, setting up, and using SkypeIn

▶ Redeeming vouchers for services

* *

*I*f people the world over were glued to their PCs, talking over Skype would be a cinch. However, people are constantly moving about and are often closer to a cell phone or a regular landline than to a computer with access to the Internet. Skype offers two kinds of services, SkypeIn and SkypeOut, to extend Skype's reach to ordinary phones. This chapter is your guide that explains what these services are, how to set them up, and how to use them.

Connecting Skype with Regular Phone Lines

The point of SkypeOut and SkypeIn is to give you seamless connectivity to all your contacts, associates, friends, and relatives, regardless of whether they are using the Internet or are using regular phone lines. These services have nice features. It is easy to dial out, and you get a record of your SkypeOut calls. SkypeIn comes with a complimentary Voicemail service, eliminating the need to be tethered to your computer. Skype Voicemail can be used for any incoming Skype calls, not just the SkypeIn ones.

You're not required to use SkypeOut or SkypeIn. When you do need to use it, though, you can have it running in a matter of minutes and at very low cost.

Understanding SkypeOut

The central concept of SkypeOut is simple. If you're on Skype, you should have the ability to speak to anyone, whether the person is on a computer attached to the Internet, or on a landline or mobile phone. Also, the process of connecting to someone on a telephone should be just as simple as skyping to a person on another computer. SkypeOut is the service that seamlessly bridges Skype users on the Internet to regular phones.

Be aware that Skype is not set up to make emergency calls such as 911. Skype is not a replacement for your ordinary phone and *cannot* be used for emergency calling.

SkypeOut pricing

A PC-to-PC connection over Skype is always free. A PC-to-phone connection entails an extra service (SkypeOut) and *may* entail a fee for its use. As of this writing, SkypeOut calls made from locations in North America to phones in North America were free from May 15th through December 31st, 2006. You can check the Skype Web site to see whether the free period has been extended. By the way, there is absolutely *no registration or signup* for the free service. Just get on Skype and use it!

When a fee is attached to a call, you are told what the per-minute charges are before you make the call. To make a call, you must first buy Skype credit. Prepayments are made in $10 increments. The next section explains how to set up your SkypeOut credit.

Setting up SkypeOut

To buy Skype credit, you need to have two important things in place:

- ✔ **A Skype password.**
- ✔ **A valid email account:** The email account that you supply is the one associated with your Skype Name (see Chapter 2 for details about establishing your Skype Name).

Without these items, you cannot buy Skype credit or make use of services such as SkypeIn or Voicemail.

Skype user profile: Christel

Christel, a young woman living and working in the French Alps, loves to travel around the world on her vacations. She visits the United States regularly and has made many friends all over the country (one of this book's authors is one of them). She tries to keep up with everyone through email, but prefers direct contact. She also wants to improve her English. She has become a veteran skyper, chatting and conferencing with all her new friends. Many of the people she has befriended during her travels have not quite moved into the information age and are still using plain old telephones, or maybe a cell phone.

Christel decides to purchase some SkypeOut minutes. The cost of a SkypeOut call from France to the United States, to a landline telephone, is about two cents a minute. Calling to a mobile phone is more — about 20 cents a minute. However, using SkypeOut gives Christel some advantages. For example, if she's calling someone in a mountainous area, cell phone service may be nonexistent or filled with "dead" zones, but the SkypeOut service works fine. Internet telephony has a consistently better voice quality than cell phones. Skyping out to a regular telephone is both inexpensive and clear sounding. It gives Christel flexibility because she can call out from any computer connected to the Internet.

Another advantage to Christel is that she can buy ten dollars' worth of SkypeOut minutes at a time. Because this is all you can buy until you use some of those minutes, Christel can easily budget her international calls so that she doesn't incur endless charges; she just uses up her "calling currency" and then decides whether to purchase more immediately or hold off for a while.

You can purchase SkypeOut credit while running your Skype application or by going to the Skype Web site (www.skype.com) from your browser. Here, we show you how to do it from your Skype program. Although the process is entirely equivalent, the screens for purchasing credit directly through the Skype application differ from the way they appear if you go directly to the Web site from your browser to purchase credit.

Purchasing credit through your Skype application

When you click the My Account panel of the main Skype window (see Figure 8-1), it expands to reveal various links, one of which is Buy Skype Credit. Depending on what Skype services you already have in place, you may see other links, including Voicemail and SkypeIn. To buy Skype credit, follow these steps:

1. **Click the Buy Skype Credit link.**

   You should momentarily see a padlock display (see Figure 8-2), signifying that you are securely connecting to the Skype site.

2. **Fill in the requested billing information.**

   You are asked to fill out some basic information (see Figure 8-3) and method of payment (which can be PayPal, Visa, MasterCard, or other forms of payments, such as a wire transfer).

- **Respond to security confirmation:** As a security protection, Skype emails a confirmation code and asks you to insert this code in a confirmation email screen (see Figure 8-4).

- **Receive feedback for your order:** Skype displays a notification that your order is being processed (see Figure 8-5) and sends you a confirmation by email.

Figure 8-1:
Buy Skype
credit from
your Skype
application.

The price is right

SkypeOut calls vary in price. From at least May 15 through the end of December 2006, SkypeOut calls originating in the U.S. and Canada made to phones in the U.S. and Canada were completely free. As of this writing, we're hoping that this deal will be extended. Besides this campaign, Skype keeps running other SkypeOut campaigns for different countries, such as Skype Gift Days and others.

Always check Skype.com for current details; there may be a campaign running just for your destination.

When Skype does charge for international calls, the cost is based on a per-minute fee. The most popular destinations have one unified rate, often referred to as the "SkypeOut Global Rate." As this book was written, the per-minute charge was € 0.017 (euros), which is approximately the same as U.S. $0.021 or £0.012 (pounds).

The countries included in the SkypeOut Global Rate are as follows: Argentina (Buenos Aires), Australia, Austria, Belgium, Canada, Canada (mobiles), Chile, China (Beijing, Guanzhou, Shanghai, Shenzhen), China (mobiles), Denmark, Estonia, France, Germany, Greece, Hong Kong, Hong Kong (mobiles), Ireland, Italy, Mexico (Mexico City, Monterrey), Netherlands, New Zealand, Norway, Poland (Poland, Gdansk, Warsaw), Portugal, Russia (Moscow, St. Petersburg), Singapore, Singapore (mobiles), South Korea, Spain, Sweden, Switzerland, Taiwan (Taipei), the United Kingdom, the United States (except Alaska and Hawaii) and the United States (for mobile phones).

Countries with "mobiles" in the SkypeOut Global Rate are charged the same 2.1 cents a minute that it costs to make a SkypeOut call to a land-line.

Calls to other countries have different individual rates. You can find these rates on the Skype Web site (www.skype.com). We also post a quick link to the Skype page on our book Web site (www.skype4dummies.com).

If your billing address is in the European Union, a value-added tax surcharge may be applied when you buy Skype credit.

Figure 8-2:
Connect
securely to
Skype.

Figure 8-3:
In the Buy
Skype Credit
form, fill in
your basic
billing
information.

Figure 8-4:
Skype
sends you a
confirmation
code via
email and
asks you to
enter it
here.

Figure 8-5:
"Your order
is received
and being
processed."

As mentioned earlier in this chapter, you can buy Skype credit only in increments of $10 at a time and you cannot buy additional Skype credit immediately after a purchase without actually using some of your Skype credit. This limitation is a fraud-protection mechanism designed to prevent computers or hackers from artificially racking up free minutes. After you've established yourself as a proper user, you get more power on the system.

Using SkypeOut

Using SkypeOut is truly simple. Just enter the phone number along with area code into the input text line near the bottom of your main Skype window; then, press Enter or click the green call button to start the call.

When you enter a phone number, SkypeOut doesn't mind whether you use hyphens or spaces. It even handles parentheses around the area code.

Dialing international calls

To make an international call, precede the phone number with a plus sign (+) followed by the country code and the area code and local number. To find the country code, use the SkypeOut Dialing Wizard. On the Windows platform, you open this wizard by choosing Tools⇨Call Forwarding and then clicking the <u>Get Help Entering Phone Numbers</u> link (see Figure 8-6).

On the Macintosh platform, you may have to add a plus (+) symbol, followed by the country code, the area code, and then the seven-digit phone number (even for local numbers). In the United States, the country code is 1. So, for example, to call toll-free information, you enter **+1 800 555 1212**.

Figure 8-6:
You find the link for accessing the SkypeOut Dialing Wizard in your Skype Options panel.

The link for the SkypeOut Dialing Wizard does not appear until you have bought SkypeOut credit.

When you click this link, you are taken to a Web page (see Figure 8-7) that lets you choose which country you want and assists you in entering the number correctly.

Figure 8-7:
The
SkypeOut
Dialing
Wizard
helps you
construct
the
telephone
number for
an
international
call.

Macintosh users have a slick feature called a Widget that enables Skype users with Skype credit to make SkypeOut calls. It incorporates the long-distance calling wizard, so you don't even need to go to a Web page to construct the phone number.

At the time you make your SkypeOut call, you know immediately what rate you are being charged. Skype alerts you to the calling rate by displaying a little caption that might say "United Kingdom 0.0253 $ per minute" (see Figure 8-8). Very handy if you are watching your budget.

The number used in this example is a purely fictitious number. If you use a number that doesn't exist or is invalid, Skype warns you (see Figure 8-9).

As a particularly nice feature of SkypeOut, when you make SkypeOut calls, Skype stores or caches the numbers even if they are not part of your Skype Contacts list. The next time you start dialing a SkypeOut number, the portion of the number you dial is matched against your Skype contacts and previously called SkypeOut numbers from the current session. For example, if you start typing the 212 area code, all your 212 numbers pop up to the top of the Contacts list. If you see your SkypeOut number in the displayed list, you can click it and call the number. This is great if you frequently make international calls to the same number. It is a time saver, and you won't risk making errors when you try to dial a long number.

Figure 8-8:
You are
alerted to
SkypeOut
calling rates
as you make
your call.

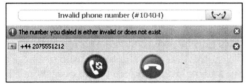

Figure 8-9:
Skype
warns you
of invalid or
nonexistent
numbers.

Navigating touch-tone and voicemail systems

Some of the SkypeOut calls you make may be to automated services that
require you to press the touch-tone keypads of your phone. Because you're
calling from a computer, what do you do when the automated voice says,
"Press 1 for Yes and 2 for No"? No problem! Skype provides you with a virtual
keypad called a dial pad (see Figure 8-10) found on the Dial tab. If your Dial
tab is not visible in the main Skype window, choose View➪View Dial from the
Skype main menu.

Figure 8-10:
Use the
Skype Dial
tab to
generate
touch tones.

To make dialing on a SkypeOut call even easier, you can copy a phone number residing in any document, such as an email, and paste it directly into the Skype Dial tab. Sometimes phone numbers have parentheses surrounding the area code. These will not cause you any problems.

SkypeOut lets you call vanity numbers, which are phone numbers that have a matching word or acronym on the dial pad of a touchtone phone. For example, to order tax forms from the Internal Revenue Service, you can call 1-800-TAX-FORM (1-800-829-3676). On the Skype Dial tab, you can enter **1-800-TAX-FORM** and click the green call button. Skype then places a SkypeOut call to the phone number 1-800-829-3676. The next time you need a tax form, you know whom to call!

Keeping track of your SkypeOut accounting information

Another nice feature of SkypeOut is that you can easily review all your SkypeOut calls in a clearly organized Web-based table (see Figure 8-11).

The steps involved in retrieving your accounting information are as follows:

1. **Go to www.skype.com and click the <u>Sign In</u> link.**

   You should see this link at the top of the page.

2. **In the text input field, enter your Skype Name and password and click Secure Sign In.**

   Your Account Overview page appears.

3. **Click the <u>Calls & SMS history</u> link.**

   You can view a list of your latest calls and SMS messages (see Figure 8-11).

My call list

<u>Summary</u> | Details

« April 2006		May 2006		June 2006 »		
« Previous		Page 2 (11 to 20 of 27 calls)			Next »	
Date	Number	Destination	Rate/min.	Duration	Price	
May 15 22:23	+12141111432	USA	0.000	09:23	0.000	
May 13 06:33	+18882113832	USA-Toll Free	0.000	02:20	0.000	
May 11 13:50	+99001114111391438		0.000	19:04	0.000	
May 11 10:37	+99001113881888215		0.000	00:00	0.000	
May 11 10:34	+99001114311121422		0.000	00:11	0.000	
May 11 08:59	+13188821000	USA	$ 0.021	00:12	$ 0.021	
May 09 18:20	+13111121113	USA	$ 0.021	20:23	$ 0.441	
May 08 21:26	+12188828863	USA	$ 0.021	01:04	$ 0.042	
May 08 21:24	+9411131113	Sri Lanka	$ 0.148	00:00	0.000	
May 07 00:16	+12133323337	USA	€ 0,017	00:32	€ 0,017	

« Previous	Page 2 (11 to 20 of 27 calls)	Next »

<u>Export</u> | Show 10 ▾ numbers on each page

Figure 8-11:
A Skype call
list.

You should be aware that SkypeOut credit needs to be kept "alive." If no SkypeOut activity occurs for 180 days, any of your remaining credit is wiped out. This does not mean that you must be using up your credit; it just means that you need to have some level of activity during any 180-day interval.

Staying in Touch Through SkypeIn

SkypeIn is unique tool that makes it easier, and cheaper, for customers and relatives to contact you because

- ✔ Calls to a SkypeIn number can be made from any landline or cell phone.
- ✔ SkypeIn numbers are based in any locality around the world.
- ✔ You can travel anywhere and be reached through your SkypeIn number.
- ✔ SkypeIn puts computer-phobic relatives and friends at ease. They can use an ordinary telephone to skype you (and they don't even know how technically proficient they are!).

Skype user profile: Lindie

Lindie, a young entrepreneur, travels nonstop giving workshops for her educational software company based in California. Because she's a moving target, her customers sometimes have difficulty reaching her when they have a question. Giving out her cell phone number is one option, but calls may be costly for teachers on a tight budget, and likewise for Lindie, who is trying to build a business in a competitive world.

To give her company a competitive edge, Lindie set up a SkypeIn account in her busiest location, California. Now teachers and school administrators throughout California can contact her wherever she is working. For some it requires a local call; others incur a toll charge, but it's still cheaper for them to call within the state than outside it. Because Lindie is almost always in front of, in the vicinity of, or toting a computer, SkypeIn is an easy choice. She also needs to be in front of a computer to answer the many technical, logistic, and curriculum-based questions that she fields. Leveraging Internet telephony to create a more valuable resource for customers and at a lower cost is the strategy Lindie put in place, and it's paying off.

Understanding SkypeIn

With SkypeIn, people in a given region or country can call you by dialing into a local phone number that you set up through Skype. It is just a local call for them, but they can reach you anytime and anywhere in the world you are, as long as you are logged onto Skype. When you get SkypeIn, you also get free Skype Voicemail, so if you're not connected to Skype when people try to reach you, you can receive messages.

When people call you using SkypeIn, they don't pay for anything other than the normal charges for calling your local SkypeIn number. If you set up a SkypeIn number that is based in London (with an area code of 207 and country code +44), anyone in London can dial the local number you give them. They pay only the cost of a local call. You can then be anywhere in the world — Madrid, Paris, Nashville, and so on. Anyone can call you from a regular telephone and can talk for hours, and all the caller pays is the cost of the local call in the London. The telephone companies are happy because they get to charge for a call. Skype is happy because you just rented a SkypeIn number for the year. Your caller is happy to talk as long as he or she wants, very inexpensively.

Now if your friend Ian from Edinburgh, Scotland, tries to reach you using your SkypeIn number based in London, he pays whatever it costs to place a call from Edinburgh to London. The cost from Edinburgh to London is not a local call, but if you are currently visiting Melbourne, Australia, it's certainly much less expensive than calling to Australia.

It's really that simple.

SkypeIn phone numbers can be acquired for Australia, Brazil, China, Denmark, Estonia, Finland, France, Germany, Hong Kong S.A.R., Japan, Poland, Sweden, Switzerland, the United Kingdom, and United States. This list of countries is expanding, as are individual area codes with the countries, so check www. skype.com from time to time for updates.

Buying SkypeIn

The process of buying SkypeIn works much the same way as buying SkypeOut credit:

1. **Go to www.skype.com and click the <u>Sign In</u> link.**

   You should see this link at the top of the page.

2. **In the text input field, enter your Skype Name and password and click Secure Sign In.**

   The My Skype Account Overview page opens, which allows you to buy all the different Skype services: SkypeIn, SkypeOut, and Skype Voicemail. This page also provides account information settings that you may have, including details on your SkypeOut calls.

3. **Directly under the topic SkypeIn, within the Skype and Ordinary Phones section, click the Buy Now button.**

   There are several Buy Now buttons. Make sure that you click the correct one immediately underneath SkypeIn!

   In the Payments section of your My Skype Account Overview page, you can choose a form of currency. If the currency is set to euros and your currency should be U.S. dollars, you can change it by clicking the <u>Change</u> link, choosing the currency type, selecting the appropriate radio button, and then clicking the Change Currency button.

   You can buy SkypeIn credit with any of about 15 different currencies. The My Skype Account Overview page is divided into various sections: Call Phones within the U.S. and Canada for Free; Skype and Ordinary Phones; Account Settings; and Payments. Within the Payments section, you can set your preferred currency. Make this choice before you buy SkypeIn credit or any other Skype credit, because if you change your currency, you will incur an exchange fee on your available Skype credit.

   You set the currency by clicking the <u>Change</u> link in the Payments section.

4. **Select the country from which your Skype phone originates.**

   You can pick from at least 14 countries: Australia, Brazil, Denmark, Estonia, Finland, France, Germany, Hong Kong S.A.R. China, Japan, Poland, Sweden, Switzerland, the United Kingdom, and the United States.

Clicking the country flag or name takes you to a new page.

5. **Select the area code and choose a local phone number from a list of available numbers within the area code.**

   Depending on the country you choose, you may or may not be prompted to select an area code (some countries have only one area code for the whole country). Skype generates a list of available numbers for you to pick from.

   Although Skype provides you with a computer-generated list of numbers, you can pick a pattern, and Skype sees whether it is available. The pattern can include letters and the asterisk (*) wildcard character. As a pattern, you can enter *****Mary** and you might get numbers like 367-6279, 123-679, or 333-6279.

   When you buy SkypeIn, Voicemail is included at no extra charge with your SkypeIn subscription.

6. **Click Buy Selected Number.**

   An invoicing page appears, and you can choose between a 3- or 12-month subscription. You're also prompted to provide billing information, including your name and address.

7. **Select your payment method.**

   Payment options include PayPal, MasterCard, Visa, regular bank transfer, Moneybookers, and JCB.

8. **Verify the accuracy of your order, complete your payment information, and click Submit.**

   If you are charging your SkypeIn credit to a credit card, you need to enter your credit card number and expiration date. Some countries may require you to pay a Value Added Tax, or VAT.

   When payment is complete, you see a confirmation page with your order number and order details. It's always a good idea to print this confirmation page.

For each SkypeIn phone number you get, you pay a set fee for either a three-month period or a full year. You can have multiple SkypeIn numbers but you're currently limited to 10 such numbers. Suppose you work for an aerospace firm with offices in San Jose, California, and Denver, Colorado. You may be doing a lot of work in Washington, D.C. You can get SkypeIn numbers for the area codes 408, 303, and 202. This way, your associates in those same area codes can reach you without incurring charges no matter where you are. You might be attending a conference in Toronto, Canada, or visiting family in North Carolina.

SkypeIn is a practical service to have. If you have a small business and want to establish an international presence, SkypeIn is a great way to leverage your resources.

When you list your SkypeIn number on your business card or company letter-head, you can identify it as a tie-line number (that is, a phone number that you set up through Skype), but be sure to indicate the time zone people are calling into. Otherwise, you may be getting calls many hours earlier or later than you are prepared to receive them!

Redeeming Skype Credit from a Voucher

Many third-party products (for example, most of the Skype-certified phones and audio-related devices described in Chapter 11) include vouchers for services such as SkypeIn, SkypeOut, or Voicemail. If you have a voucher, you can redeem it by following these steps.

1. **Go to the following Skype Web page (see Figure 8-12):**

   https://secure.skype.com/store/voucher/redeem.html

   Enter your Skype Name and password. If you are already signed in (you may have recently looked at your My Skype Account Overview page, which requires that you sign in), your Skype Name automatically appears in the form and you are not asked to provide a password. No sense in logging in when you're already logged in!

   The Skype Web site is constantly undergoing changes, so the URL for this page may change. Check this book's Web site (www.skype4dummies.com) from time to time for updates.

2. **Enter your voucher number in the Enter Voucher Details text input box.**

   You don't have to specify the kind of Skype credit you are redeeming. From your voucher number, Skype can automatically tell whether you are redeeming SkypeIn credit, SkypeOut credit, Skype Voicemail, or any other type of Skype-related service or product.

3. **Select the I Agree to the Skype Terms of Service check box and then click the Redeem Voucher button.**

Depending on what you are redeeming, you can get an instant confirmation (such as for Voicemail or SkypeOut, or to extend existing SkypeIn service) that your voucher was successfully redeemed (see Figure 8-13). Other types of services may require additional processing. For example, you may be purchasing SkypeIn for the first time. In this case, you need to specify the country code, area code, and so on. Rather than pay with a credit card or something similar, you can use your voucher.

Figure 8-12:
Redeeming a voucher.

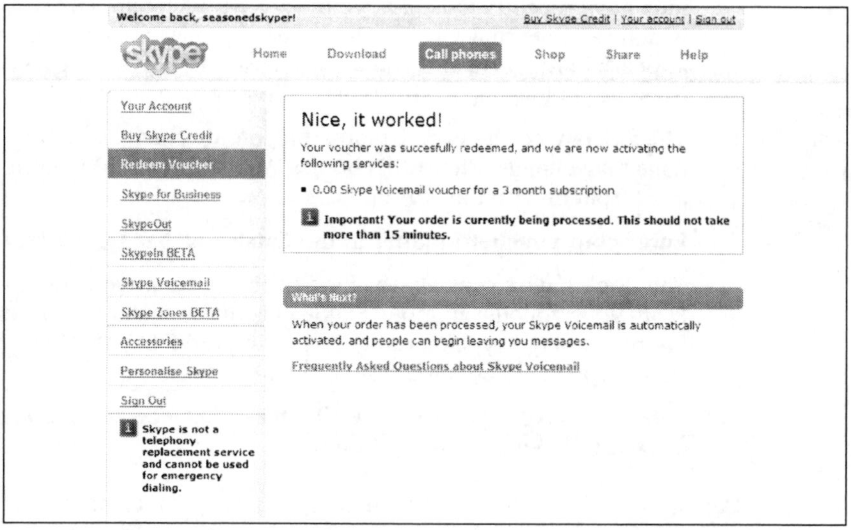

Figure 8-13:
Receiving confirmation that a voucher is successfully redeemed.

Part III
Calling All Seasoned Skypers

The 5th Wave   By Rich Tennant

"I had a little trouble with the automatic video tracking camera, so during the video conference, before speaking, say 'Here Rollo!' and wait for Rollo to get his paws up on your knees before beginning to speak."

In this part . . .

This part is jam-packed with the basics on tools for both enjoyment and business productivity, such as voicemail, call forwarding, text messaging, conferencing, and sending files in the blink of an eye. In addition, find out how to record Skype conversations for fun, posterity, and because you can't ever remember what anyone says.

Then it's time for some Skype shopping! Also in this part, you can browse through descriptions, scenarios, and pictures of many cool gadgets and software add-ons that work with Skype. Discover a webcam that follows you around the room, and find out how to make your computer listen and obey when you speak and how to skype on the run.

Chapter 9

Managing Your Messages

In This Chapter

▶ Getting the most from basic Skype messaging

▶ Doing the math: Skype + Outlook = Skylook

▶ Checking out Pamela; she's not just another pretty voice

Skype can follow you just about anywhere on the planet. Suppose you're flying on a jet with an airline that has Internet access, or perhaps you're hiking up a trail on Mt. Everest and have a wireless connection to the Internet. Even in these far reaches, anyone around the world (anyone appropriately equipped) can reach you. You simply pack a U3 Smart Drive or a Wi-Fi phone and tap into the Internet wherever it's available.

Having this connection on hand doesn't mean that you're available or that you actually want to answer every call, all the time. But having voicemail and even video mail allows other people to connect to you without requiring your physical presence to take every incoming call.

In this chapter, we tell you about various methods for handling the Skype calls that you can't answer in the traditional way. Skype services do include some basic voicemail, call forwarding, and text messaging. You can also add on to or integrate other software with Skype to accept voice messages, record contact-specific personal greetings and conversations, and even schedule Skype calls with your contacts. In particular, we describe how to make Skype collaborate with Microsoft Outlook to organize and store voice messages, as well as how to use Pamela, one of the earliest Skype add-on programs for voice messaging.

Staying Connected via Skype

Voicemail is nothing new, and Skype comes with basic voicemail capabilities as an optional feature that you can choose to purchase. If you decide to buy the SkypeIn service (see Chapter 8 for information), your voicemail costs change from fee to free! That's because Skype Voicemail is bundled in as a free service when you get a SkypeIn number.

Skype offers two other familiar features for helping you manage your messages: call forwarding and SMS (Short Message System) messaging, which is a standard used for text messaging. You may incur associated fees if you forward calls to a landline or cell phone, and fees for sending text messages depend on the message length and the number of recipients.

But the best part about these three messaging features is that they're easy to use. The next sections show you how.

Taking messages with Skype Voicemail

Skype supports an easy-to-use voicemail system. One convenient benefit of using Skype Voicemail is that your audio messages — both your greeting and the incoming messages people leave you — are not stored on your computer but instead are stored remotely on Skype's central servers. As a result, you don't have to worry about missing calls when you turn off your computer.

Before you can use Skype Voicemail, you have a bit of setup to do. The following steps show you how to start your setup and record the Voicemail greeting that your callers will hear:

1. **Start Skype and choose Tools⇨Options from the main menu.**

   The Skype Options dialog box appears.

2. **In the list on the left side of the Options dialog box, click Call Forwarding & Voicemail, as shown in Figure 9-1.**

   The right portion of the Options dialog box changes to show the settings related to these two services.

3. **In the Voicemail section on the right, select the Send Unanswered Calls to Skype Voicemail check box to activate your voicemail.**

   Below the Voicemail section of the dialog box, you find the Welcome Message section containing the three buttons you'll use to record and play back your Voicemail greeting.

4. **In the Welcome Message section of the dialog box, click the Record button (the button with the triangle) and speak the Voicemail greeting you want callers to hear (see Figure 9-1).**

   Your greeting or welcome message may be up to 60 seconds in length. If you have a SkypeIn number, don't forget to use part of that time in your welcome message to tell your callers to leave their phone number.

5. **Click the Stop button (the button with the dot) when you're finished recording.**

6. **Click the Replay button (the button with the arrow) to listen to the message you just recorded.**

   If you don't like what you hear, you can return to Step 4 and re-record your greeting.

7. **When you're satisfied that you recorded the perfect message, click Save.**

Skype Voicemail has an advanced setting that you can use to avoid being interrupted by additional calls if you're already speaking on a call. To select this setting, click the Advanced Settings link in the Voicemail section of the Options dialog box (refer to Figure 9-1) and select the I Am Already in a Call check box in the resulting Advanced Voicemail Settings dialog box (see Figure 9-2).

Figure 9-1:
Adjust your Voicemail settings from the Skype Options dialog box.

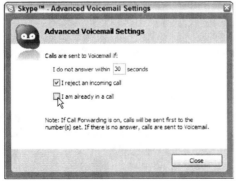

Figure 9-2:
Tell Skype to automatically record messages whenever you're speaking on a call.

You also find settings for other situations in which calls are sent to your Voicemail, as follows:

✔ **When you don't answer:** In the appropriate text box, enter a number (of seconds) that tells Skype how long to wait for you to answer the phone.

✔ **When you reject an incoming call:** Select the appropriately named check box if you want to send rejected calls to Voicemail.

You can easily sign up for or renew your Voicemail service by going to your Skype account log-in page and clicking the Buy Now button under Skype Voicemail. If you are already using the Skype Voicemail service, you should see an Extend button instead of Buy Now. The URL for reaching your login page is generally `https://secure.skype.com/store/myaccount/overview.html`.

Forwarding calls when you can't answer

Skype's call forwarding feature is very simple to set up and operate. It works like this: If someone calls you on Skype and you're not able to answer, the call is routed to the first Skype ID or SkypeOut number in your call-forward list. If the call receives no answer there, it is routed to the next number on your call-forward list.

You can call forward to as many as three Skype IDs or SkypeOut numbers. Skype-to-Skype forwarding is always free. Forwarding to a landline or to a cell phone incurs the regular SkypeOut charges and requires that your account have SkypeOut credit. The account that does the skyping out pays for the call, which makes sense. Suppose you're in the United States and, from your PC, you skype someone located in France. If that person has enabled call forwarding to his or her cell phone, then he or she (and not you) incurs the SkypeOut charges.

For basic call forwarding, follow these steps:

1. **Start Skype and choose Tools⇨Options from the main menu.**

   The Skype Options dialog box appears.

2. **Click Call Forwarding & Voicemail.**

3. **Select (check) the Forward Calls When I Am Not on Skype check box, enter the Skype name you want your calls forwarded to in the Enter Phone Number combo box, and click Save.**

For call forwarding to more than one Skype name or SkypeOut number, follow these steps:

1. **Start Skype and choose Tools⇨Options from the main menu.**

   The Skype Options dialog box appears.

2. **Click Call Forwarding & Voicemail.**

3. **Click the <u>Advanced Settings</u> link, select (check) the Forward Calls When I Am Not on Skype check box, and enter the Skype names you want your calls forwarded to in the Enter Phone Number combo box.**

4. **Enter the additional Skype names or SkypeOut numbers (see Figure 9-3) as appropriate in the remaining Enter Phone Number combo boxes and click Save.**

   If you are uncertain how to enter a SkypeOut number, click the <u>Get Help Entering Phone Numbers</u> link, which takes you to the Web page at www. skype.com/products/skypeout/rates/dialing.html. This page provides assistance on correctly entering SkypeOut numbers, including looking up the appropriate country codes.

   You must have Skype credit to forward a call to a SkypeOut number (see Chapter 8 for information about Skype credit).

Interestingly, Skype Voicemail does not record forwarded messages. Systems such as Pamela (described later in the chapter) do, and they add some other very cool features. Skype Voicemail and call forwarding offer basic functionality, but other messaging features in Skype (such as SMS messaging) and add-on products that work with Skype (such as Skylook and Pamela) offer you more.

Figure 9-3:
Set up a call-forward list through the <u>Advanced Settings</u> link.

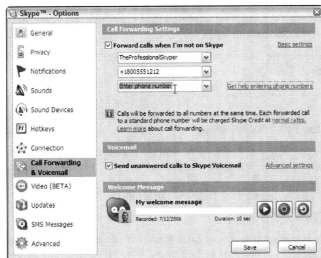

Making it short and sweet with SMS messaging

The blossoming of technologies for communication shows that people crave constant contact with each other. Nowhere is this trend more evident than in the popularity of text messaging, which is a communication method that teenagers use frequently because it's fun, easy, and (perhaps) even addictive.

Many people find that text messaging is practical, even though it has these two drawbacks:

- **Limited screen size for message viewing:** Any text message sent must be viewable from a tiny cell phone screen.

- **Tedious and time-consuming creation:** Typing a message — beyond a short sentence or two — takes more patience and time than most people (especially teenagers) are willing to invest.

To help work around these text messaging limitations, a communications standard — the Short Message System, or SMS messaging — exists for a variety of mobile phones and other devices such as Pocket PCs and GSM (Global System for Mobile Communications) phones. And when you're running Skype, the circle of devices offering SMS text messaging capability even includes your desktop computer.

Sending SMS messages via Skype involves these simple steps:

1. **In your Skype menu, choose Tools⇔Send SMS Message.**

   A window pops up, prompting you to select the recipients you want to receive your message. Recipients need not be Skype Names. You can also enter phone numbers.

2. **Add the Skype Names or phone numbers of your message recipients and click OK.**

   An SMS message window appears.

3. **Enter your text message in the Skype SMS window (see Figure 9-4) and click the Send SMS button.**

   The Skype Names or phone numbers of the recipient(s) appear in the title bar of the Skype SMS window.

That's all there is to sending the text message with Skype, but here are some points to keep in mind about this service:

- **SMS messaging has a cost attached:** When you use SMS messaging, you pay a small fee that's based on two things: the length of your message and the number of recipients getting the message.

✔ **Charges apply only to messages delivered within 24 hours:** When you send a message, you are charged only if the message reaches the person within 24 hours. If the recipient's GSM phone battery is dead or his or her phone is off, you get a refund.

✔ **Watching your message length is a good idea:** The maximum size for any SMS message is 160 characters. So if you send a message with 167 characters, it is split into two SMS messages with 160 characters in the first and 7 in the second. In this situation, you are sending two messages instead of one and you'll be charged accordingly. (It pays to study up on the chat abbreviations presented in Chapter 6!)

Figure 9-4:
Sending
your SMS
recipients
a text
message.

Bridging Skype with Your Outlook Contacts

Skype is all about connecting with people. And when you use SkypeOut (see Chapter 8), you are only a click away from making that connection. But here's a problem: All the information about the people you want to connect with (your contacts) may live on a separate island known as Outlook.

Wouldn't it be nice if Skype could just take a peek at your Outlook contacts, see their phone numbers, and skype out to them by clicking on their phone numbers from within Skype? Starting with Skype version 2.5 and later, you can make this happen by following two simple steps:

1. **Open both your Microsoft Outlook program and Skype (on Windows) so that the two programs are running at the same time.**

2. **Choose View➪View Outlook Contacts from the Skype menu.**

Don't confuse the Outlook program with Outlook Express. Skype and Outlook work together very happily, but you won't get Outlook Express to share information with Skype.

What do you get when you cross Skype with Outlook?

Skype is a communications engine that enables you to talk, write, watch video images, listen, and chat instantly with a variety of contacts. When you use Skype, you can reach skypers, nonskypers, and text messagers, all from one program. Outlook is also a communications engine that helps you manage communications with your contacts as well as information about your activities and your schedule. With Outlook, you compose, receive, and archive emails, make appointments, schedule your time, and share your calendars and folders.

Both Skype and Outlook are amazing programs that offer access to a broad circle of friends, associates, and family. So it stands to reason that both programs work with much of the same information (contact information, that is). Consider the following strengths of each program:

- ✔ **Skype lets you connect with your contacts by exchanging information through the Web** and extends that outreach with sound (voice calls), sight (video), live text messaging, and file transfer.

- ✔ **Outlook connects you with your contacts by moving information around through various networks,** extends its reach to organizing and scheduling your activities, and even gives you a way to send information to hundreds (or thousands) of people at one time.

Without any integration between the contact information stored in Skype and Outlook, you can quickly get caught up in switching back and forth between the two programs to replicate and update the same information in each one. This inefficient jockeying is the quintessential example of electronic pencil pushing at its worst. So to alleviate the annoying repetition and duplication, Skylook (a software plugin used with Outlook) was born.

Plugging in and setting up Skylook

You can get the Skylook plugin to use with your Outlook program by doing the following:

1. **Install Skype 1.2 or later, and install Outlook 2000 or later.**

   Skylook is not an independent program; it is a plugin. You need to have both Skype and Outlook properly installed before downloading and installing Skylook.

2. **Quit Outlook.**

   Outlook must be closed in order for Skylook to be installed. Skype can remain open.

3. **On the Internet, navigate to `www.skylook.biz`.**

4. **Click Download.**

   You are prompted to choose a place to save the Skylook installer file. Click Save to place the file in a default folder on your hard drive. You can choose any location, as long as you can locate the file on a program such as the Windows Explorer. Jot down the path to the folder where you save the Skylook installer file.

   The basic Skylook account is free. Additional features require a small incremental fee.

5. **Launch the Skylook install wizard by choosing Start➪All Programs➪ Accessories➪Windows Explorer; then, navigate to the installer file you just saved and double-click the file.**

6. **To install Skylook, you must accept the terms of the End User License Agreement. Click I Accept the Terms in the License Agreement and then click Next.**

   Before proceeding with the installation, make sure that you close the Microsoft Outlook application if it is running. The Skylook installer program does some behind-the-scenes hocus pocus so that the next time you open Outlook, Skylook kicks into action and prompts you with a series of setup questions.

7. **When you finish the Skylook installation, launch Outlook.**

   Magically, Skylook is integrated into Outlook, but it still needs some configuration. The Skype security screen (shown in Figure 9-5) appears on your desktop. This same screen appears when any program that wants to use Skype opens. To continue, you must officially introduce the two programs and agree to let Skylook and Skype work together.

8. **If you use Outlook all the time, select the Allow This Program to Use Skype option and click OK.**

   Choosing this option eliminates the security reminder every time you launch Outlook, which is a good idea if you're a daily Outlook user.

Figure 9-5:
The Skype
API security
screen.

9. **The Skylook setup wizard opens automatically.**

Follow the prompts to configure Skylook and get a feel for how Skylook works at the same time.

You may be asked to answer questions relating to the following:

- Whether to synchronize your Outlook contacts with your Skype contacts. We show you how to do so in the next section.

- Whether you want Skylook to record your conversation (either one or both sides).

- Whether you want Skylook to act as your answering machine. If you do, you can record a personalized greeting, set Skylook to automatically answer when the Skype status is Away, Not Available, or Do Not Disturb, and specify how many seconds to wait before answering.

You have the option to change any of these settings after the program is fully installed.

To use the call recording feature of Skylook, you need to have Windows Media Player 9 or later installed. Download Media Player at www.microsoft.com/ downloads.

Skylook binds Skype and Outlook tightly together. You don't sense that two separate programs are running at the same time. You see one interface, one program to open, and one environment in which to store and send messages. But Skype and Outlook are, in fact, two separate pieces of software. To truly have them work together, you need to give them a little help connecting.

Toolbar too crowded?

If you can't find your orange Skylook icon, you have too many words on your Skylook toolbar. Here's how to get rid of them.

1. **Move your mouse to the far right end of the toolbar where you see a small, triangular notch; click the notch to open your toolbar options.**

2. **Choose Tools⇨Options.**

   A Skylook Options window opens.

3. **Click the Toolbar tab and select the first check box, Hide Captions.**

4. **For the Show Contact Presence Buttons in the Toolbar option, set the number to 3 (see the following figure).**

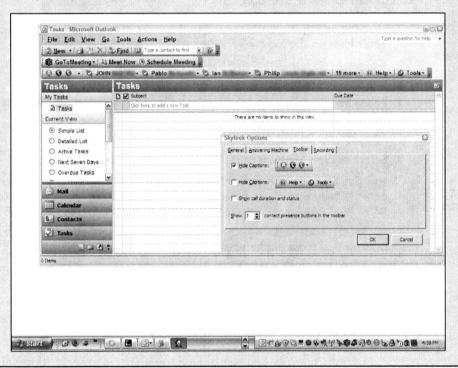

Getting Skype and Outlook in sync

Skylook has some slick synchronization tools to pull Skype contacts into Outlook and to roam through Outlook contacts to find Skype users. At times, Skylook walks you through individual connections, as it does when you create a new appointment. Veteran Outlook users make (and have made) hundreds of appointments in Outlook. When you add Skylook to Outlook, the appointment screen takes on a new Skylook toolbar, on which you can conveniently make calls on the spot and then schedule your Outlook appointments all from a single window.

Before you can initiate a Skype call within Outlook, you must synchronize your Outlook contacts with your Skype contacts. Follow these steps to synchronize contacts manually:

1. **Find the small, triangular notch to the right of the orange Skylook Tools icon on the Skylook toolbar.**

   When you click this "notch," the Skylook Options and Other Functions menu appears.

2. **Select SkySync (Synchronize/Import Contacts) to open the SkySync Wizard.**

   The SkySync Wizard opens a SkySync Summary window.

3. **Select Review Matches and/or Link Contacts Manually.**

   You also have the choice to let SkySync search the entire global Skype network to find matches between your Skype contacts and Outlook contacts. It takes a few minutes to perform this search.

4. **SkySync Wizard asks whether you want to update Outlook contact information from Skype profiles. Click Yes; then click Next.**

   Skype profiles sometimes have extra contact information such as phone numbers that can be added to an Outlook account.

5. **Select an Outlook Contact from the list that the SkySync Wizard generates.**

   The Review SkySync Recommendations window offers several choices:

   • **Link:** Connects a particular Skype contact with the Outlook contact selected.

   • **Unlink:** Severs Skype from the selected contact.

   • **View Diffs (differences):** Lets you see the profile details of each account.

   • **Update:** Updates your Outlook account with the latest profile details.

   • **Don't Update:** Leaves each profile unchanged. Sometimes Skype profiles are deliberately different from Outlook profiles. One may be business related; the other may be personal.

6. **Click Link.**

   When you click the Link button, a window opens and the SkySync Wizard asks you to enter the Skype User ID (Skype refers to this as your Skype Name; Skylook calls it your Skype User ID) of your Outlook contact in the box. If you don't know the ID, Skylook searches for it when you click Next.

7. **Select a name from your list of Skype Contacts generated by SkySync.**

8. **If you still don't see your Skype contact, click Next.**

   A search screen opens. You can add some search details, including:

   - **Full Name**

   - **Email Address:** Skype Profiles never reveal email addresses, but you can still search for a contact if you already know the email address and it is included in your friend's Skype profile.

   - **Skype User ID:** You can enter a portion of the Skype User ID if you don't remember the entire name.

     Click Next and select the Skype User ID from the list the SkySync Wizard generates.

9. **Click Finish.**

   Your Outlook contact is in sync with Skype.

Synchronizing contacts manually through the SkySync Wizard lets you scroll through contacts and decide whether to sync each one individually. However, if you want to synchronize a specific Outlook contact on the fly, here is a little shortcut:

1. **Select a name in your Outlook Contacts list.**

   The selected name appears in your Skylook toolbar between the Skype Phone Call button and the Help button.

2. **Click the arrow next to the Outlook contact that now appears in the Skylook toolbar and choose Link This Outlook Contact to a Skype Contact from the drop-down menu, as shown in Figure 9-6.**

3. **Follow the screen prompts (as in Steps 5–8 in the prior step list) to synchronize the contact with the list of Skype contacts (see Figure 9-7).**

4. **Click Finish.**

Skylook can expedite contact linking between Skype and Outlook with its SkySync feature. Rather than connect one user at a time, SkySync searches through your Outlook address book and maps it to your Skype Contacts list by linking any matches that it finds. You can even use the SkySync Wizard, shown in Figure 9-8, to run a thorough search of the global Skype network.

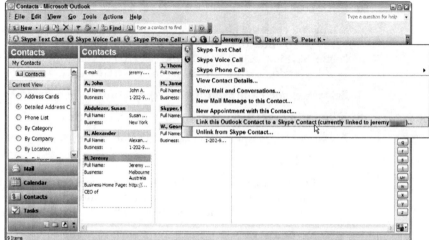

Figure 9-6:
Outlook
contact
being linked
to a Skype
Contact
name in
Skylook.

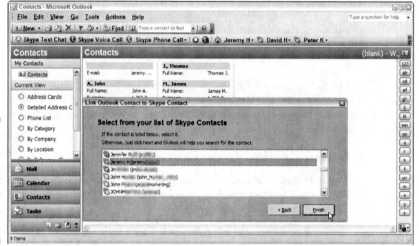

Figure 9-7:
Selecting an
Outlook
contact
from a list
of Skype
contacts.

Suppose that your company has several thousand active Outlook accounts. Synching up one contact at a time would be out of the question, but using the SkySync Wizard makes this contact mapping possible. SkySync launches automatically during the Skylook setup, but if you want to update your contact links subsequently, choose Tools⇨SkySync to open the SkySync Wizard and select Search the Global Skype Network for More Matching Skype Users.

SkySync goes through all your Outlook contacts and searches for possible matches on the Skype network. Skylook presents possible matches. For each potential match, Skylook asks you whether you want them linked. Suddenly you see that good old Bob has a Skype account, and SkySync links it to Bob's contact information in your Outlook program.

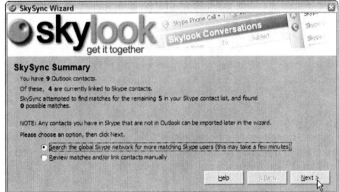

Figure 9-8: A SkySync summary of linked Outlook and Skype Contacts.

After your Outlook Contacts link up with Skype, simply select Bob from your contacts in Outlook and skype him via the Skylook plugin. To start a new Skype conversation with Bob, follow these simple steps:

1. **Select your Outlook Contact name.**

2. **Click Skype Voice Call in the Skylook toolbar.**

   Skylook double-checks your intentions and asks whether you *really* want to call your contact.

3. **If you *really* want to have a Skype conversation, click OK.**

4. **Start talking!**

Handling Voice Messaging with Skylook

A powerful feature of the Skylook plugin adds the capabilities for voice recording and messaging from within Outlook. These communication methods are not native to Outlook, whose strength is the organization of email, asynchronous messaging, and scheduling.

Including VoIP functionality (offered through Skype) in your Outlook program suddenly catapults messaging to a new level. You can not only email your contacts and invite them to a meeting but also conduct that meeting within the same program. Business collaboration becomes a live event, bringing all the parties together synchronously. When you add Skylook to Outlook, callers can now reach you directly through the messaging system you use most often.

Please leave a message . . .

Outlook users tend to keep their Inbox open all the time, expecting little alert beeps when emails come in. A system that is always open like this needs to also handle messages delivered when the intended recipient is away from the desk. Outlook archives and organizes email messages, but it has limited capacity to respond automatically. Usually, the only option is a general auto-mated email reply that answers for you when you're on vacation or at a con-ference. But because Skype adds voice capabilities to the mix, Skylook adds voicemail through the Skylook Answering Machine.

The Skylook Answering Machine records calls and stores voicemail messages in your Inbox, handling them just like Outlook handles emails. You control when the answering machine responds according to your choice of availabil-ity options that you find in your Skype program.

Following are some options you have for controlling the Skylook Answering Machine:

- ✔ **When and how to answer:** If you choose, the answering machine acti-vates when your Skype status displays Away, Not Available, or Do not Disturb. You can also choose how long to let the Skype caller ring before the answering machine picks up, even if your Web status shows you are available.

- ✔ **See who's calling before you answer:** Sometimes you just don't feel like talking. Of course, because it's a Skype call, you can see who is calling, glance at the caller's profile, do a quick search for archived messages, and then decide whether you want to pick up.

- ✔ **Listen (or not) to messages coming in:** If you decide not to pick up a call, you can hear the message as it's being recorded, or you can mute the sound if you don't want to be disturbed.

- ✔ **Work in conjunction with Skype Voicemail:** Skylook Answering Machine works even if you have Skype Voicemail service (see the section "Taking messages with Skype Voicemail," earlier in this chapter). The two don't fight each other to answer calls. Whichever picks up first takes the mes-sage. Whether a call comes in from Skype Voicemail or through Skylook, all incoming messages are stored in the Outlook Inbox if the program is open. If Skylook is not active, Skype Voicemail takes the messages.

- ✔ **Record a personal greeting for callers to hear:** Skylook has a built-in voice recorder, so a simple wizard screen offers the choice to record a personalized greeting.

On the surface, the Skylook voicemail system works just like ordinary voice-mail. Someone calls, hears a greeting, and leaves a message. But because these messages are digital and show up in the Outlook Inbox, you can sort, save, and even email them. And voice messages aren't the only kind saved; Skylook will also store missed text chat messages in your Inbox.

Using Outlook with the Skylook plugin gives your clients, friends, and family even more ways to communicate with you. Direct calls, emails, text chats, and voice messages are all available — and you are no longer unreachable.

Making WAVs or, better yet, MP3s

Another new productivity tool Skylook brings to Outlook is conversation recording and archiving. Suppose that you set up an Outlook appointment to skype Bob and discuss the details of an upcoming workshop. Rather than take notes during your conversation, you can use Skylook's tools to record, preserve, and store the conversation.

To the right of the orange Skylook Tools icon on the Skylook toolbar is a small triangular "notch" for Skylook Options and Other Functions. Clicking this icon opens a menu. Select Options and click the Recording tab; from there, you can set the recording conversations features with the following choices (see Figure 9-9):

- ✓ **To automatically record all calls.** When automatic recording is active, the recording starts as soon as you make or receive a Skype call.

- ✓ **To record only your side or both sides of the conversation.** Select either Record My Voice Only or Record All Voices. You can set this option to apply separately to Skype-to-Skype calls or to SkypeIn and SkypeOut calls.

Recording the voices of other parties may be subject to legal limitations in some jurisdictions. These limitations may require you to notify the other party that the conversation is being recorded. In some cases, recording the other party, even with notice, may not be legal. Please verify the limitations for your jurisdiction. Even though you may be recording a conversation for your own internal use, it is always good to be considerate and courteous and inform the other party that you are recording the conversation.

The Skype Call Monitor is a floating window — shown in Figure 9-10 — that pops up to let you know a call is being recorded. When you expand the Skype Call Monitor, it becomes a larger window that displays real-time information about the recording in progress. The information that is monitored includes the length of the recording (in minutes and seconds) as it happens and the size of the recorded file (in KB) as it grows. If you capture every Skype conversation, you may need a lot more hard drive space (and tons of free time to listen to all those conversations again).

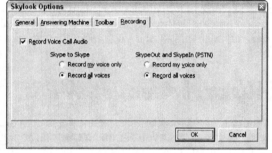

Figure 9-9:
Setting
Skylook
recording
options.

Figure 9-10:
Tracking
time and file
size with the
Skylook Call
Monitor
Window.

Suppose that you schedule a meeting (that is, a Skype conversation) with your good friend Bob in Outlook. When the meeting is about to start, a screen pops up as a reminder to make your Skype call. Click the reminder icon to open your appointment window. You then select Bob's name in the Contacts box, click the Skype Voice Call button, and you are connected over Skype to good old Bob. If you have call recording enabled, the conversation is automatically recorded and saved as an MP3 file when you log off the call. Skylook then parks the MP3 file in a Skylook Conversations folder.

Voice call recording is a full-productivity-mode feature of Skylook. This feature is available in the business, home, or academic versions of Skylook. Basic Skylook, which is free, does not offer any recording tools.

In addition to recording conversations, this full productivity mode adds contacts to your Skylook Toolbar, saves your voicemail messages to your Outlook Inbox, and creates a Skype Conversations folder to gather and organize Skype conversations.

When you have full-productivity mode enabled, Skylook presents you with choices in the Skylook Recorded the Conversation For You dialog box (see Figure 9-11). For example, you have the option of emailing the MP3 of your just-recorded conversation to Bob. This handy option saves you from rummaging around trying to find the recording. Simply click the Email a Copy To button, and the recording is on its way.

Figure 9-11:
Skylook
window for
emailing
recorded
conver-
sations

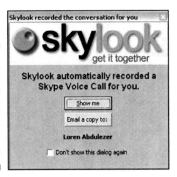

If you don't want to record a call, just click the Stop button in the Call Monitor toolbar. If you stop a recording during a Skylook call, the entire recording is deleted. If you want a portion of a conversation preserved, hang up and then reconnect to finish the part you don't want recorded. The MP3 file of your recorded portion will be tucked away in the Skylook Conversations folder.

Try using Skylook's recording feature to produce podcasts or online presentations. You can set Skylook options to make sure that you record both sides of the conversation for interviews or conferences. And if you use Skylook calling to coach, mentor, or present to a group, you can choose to record only your lecture so that you avoid capturing any noise from the others on the call.

Organizing messages

When you install the Skylook plugin, the program automatically creates a Skylook Conversations folder (see Figure 9-12) that you can access within Outlook. The Skylook Conversations folder functions just like the folders that store your email, but it contains voicemail, chats, and recorded sessions, as well.

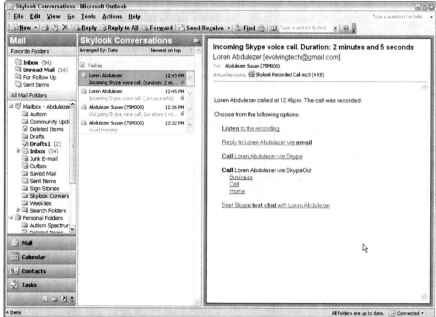

Figure 9-12:
Skylook
Conver-
sations
folder in
Outlook.

Consider these ideas for keeping the contents of your Skylook Conversations folder organized:

✔ **Treat the files as though they were emails.** Move conversations to another folder, flag them, mark them as unread, or delete them.

✔ **Organize your archived conversations** by date, category, from, for, size, subject, type, flag, email, importance, and even by *conversation*.

✔ **Review all communication in a particular thread**, or with a particular individual, to get a snapshot of an ongoing project. Go through your messages by listening, reading, and printing chats.

✔ **Share your messages by forwarding them** to friends and colleagues through email or by collecting and putting them in a public folder.

When your messages are properly sorted, you can follow up on each one more easily. You can return a Skype call, send a reply to that chat, or listen once again to a voice message or recorded conversation. You know which contacts are online because the Skype status system (the feature that shows your online Skype availability) works within Skylook. If someone's status is Online, Available, or better yet, in Skype Me mode, by all means, make that call!

Pamela Is Not Just Another Pretty Voice

Named after the Skype alias of a Skype marketer's girlfriend, Pamela is one of the first add-on programs available for Skype. Not only is Pamela (like her namesake) elegant, she's also well designed and feature rich. Pamela comes in various editions (distinguished by varying levels of functionality) and is a handy tool for managing your Skype messages. (Check out www.pamelasystems.com for how-to information on getting Pamela.) The sidebar "Getting to know Pamela" gives you a peek at Pamela's background, and the features illustrated in this chapter come from the Professional Edition.

Rather than bore you with a list of features that you can find on numerous product data sheets online, we think it is more interesting, revealing, and in keeping with the spirit of playfulness to introduce you to the real personality of Pamela.

So what can we say about Pamela? Well, Pamela is alert; she doesn't forget anything (unless you tell her to). She is very attentive. This gal has real Web savvy and she always knows whom she's talking to. Pamela is appropriately polite and punctual. But of course, she wouldn't be very interesting if she didn't have her moods.

Pamela is alert

When you work with Skype, you'll discover a tendency to start accumulating Skype contacts. You may also discover a tendency to overlook some of the information that's floating around in your Skype database, but Pamela helps you realize just how much info you have.

Getting to know Pamela

The seeds for Pamela were first sown during the infancy of Skype. With only 10,000 Skype users on the whole planet at that time, Skype was like a diamond in the rough. This situation offered a splendid combination of high potential, the ability to accommodate improvements, and the longing for someone to take and enhance the product. That someone was Dick Schiferli, the head of Pamela Systems. Over the last two years, both Skype and Pamela have evolved and continued to incorporate user-requested features.

Pamela is a well-thought-out tool for managing your messages. You'll find different levels of functionality depending on which edition of the product that you use. There's a Basic Edition that you can download for free. But rather than encourage you to go wading in that kiddy pool, we think you probably want to know (and put to use) a set of features found in the higher-end editions of Pamela. Go check out the Professional Edition; you'll be glad you did.

For example, we like to experiment and tinker, so we decided to launch Pamela with a reasonable number of Skype contacts just to see how it operates. To our surprise, while Pamela loaded, it reminded us that our friend Hannes' birthday was that day (see Figure 9-13) and prompted us to make contact. We didn't expect this kind of volunteered information, but we certainly welcomed it!

Figure 9-13:
Pamela offers a sweet birthday reminder.

Conveniently, Pamela scours your contacts list to see whose birthday is around the corner. She can also help you send out birthday greetings with the options shown in Figure 9-14.

Figure 9-14:
Birthday notification alert settings.

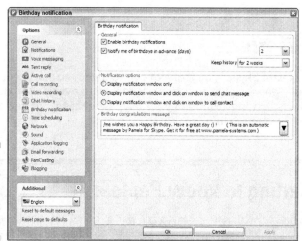

Pamela has a great memory recall

Pamela records voicemail messages while you are away, and she can also record any conversation as you're making or receiving calls. With the Professional Edition of Pamela, the number of messages you can record is limited only by the amount of available disk space. Pamela is configured to record conversations using uncompressed WAV files.

And although having large amounts of hard disk space at your disposal is getting less expensive these days, you may still feel the need to conserve. If you want to cut down the disk space used by a factor of 12 or so, you can install a couple of special files that make Pamela save sound recordings as MP3 files. You can find the instructions for making this change at `www.pamela-systems.com/download/mp3.php`.

As download links or URLs listed in this book change, we will post updated links on our Web site `www.skype4dummies.com`.

A couple of items about Pamela's message recording features are worth pointing out:

- ✔ **Pamela can automatically record all Skype conversations as you have them.** Auto recording all conversations, as well as voicemail messages, imposes two burdens on you. First, your hard drive could fill up very quickly, and second, you have to figure out how to manage all those recordings (and then do it!). Alternatively, Pamela can prompt you each time you make or receive a call over Skype and ask you then whether you want it recorded.

  If you ignore the recording prompt for 10 seconds, Pamela assumes that you are not interested (in having the conversation recorded, that is) and closes the prompt.

- ✔ **Pamela can help keep your recordings legal.** To help keep you in compliance with legal rules in various countries and geographic regions, Pamela can display an alert to let people know that the conversation is being recorded.

If you want, you can set up Pamela to record calls in stereo. That is, your voice is recorded on the left channel or speaker, and the person or people on the other end of the call are recorded on the right. This type of recording is great for podcasting because editing for clarity becomes easier. For example, you can safely eliminate discernable background noise on an interviewer's side of the call while preserving the sound on the interviewee's side.

Pamela is attentive

Imagine that during the first week of using Pamela, you record a half-dozen conversations — each about a half-hour long — with the same party. Other than by listening through each of the conversations one-by-one, how will you remember the key points and which conversation they belong to? Fear not; Pamela to the rescue!

Pamela's note-taking function (shown in Figure 9-15) allows you to record the date, time, and identity of the person you are speaking to, and to type in whatever notes you want. These notes are appended to the sound recordings so that, at a glance, important details and quick, on-the-spot summaries are at your fingertips. Joe Friday would be pleased with this software. *Just the facts, ma'am!*

Pamela knows whom she's talking to

Want something really cool and way over the top? Try making personalized greetings for any Skype contact. That's right! You can record a different greeting message for your husband, your boss, your favorite niece, and that annoying neighbor who feels obligated to gossip about everyone on the local school board. Figure 9-16 shows the Personal Options dialog box with Enable Personalized Voice Mail for This User selected (in this example, the user is NewbieSkyper).

If you think that personalization alone is cool, wait until you combine personalization with brains. We discuss call forwarding earlier in this chapter (see the section "Forwarding calls when you can't answer"). So what happens if NewbieSkyper calls SeasonedSkyper, who has call forwarding turned on, and the call is punted over to TheProfessionalSkyper running Pamela? Happily, Pamela is smart enough to know who is actually doing the calling and applies the personalized greeting for NewbieSkyper.

Figure 9-15: Getting your conversation facts in order.

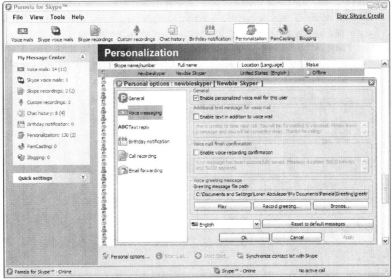

Figure 9-16:
Record a
personal
greeting for
Newbie
Skyper.

The personalization feature may be cool, but it is also practical. Suppose you are traveling and have access only to email. During your travels, Pamela may have logged plenty of messages on your home base. But also suppose that you're expecting a very important message from one of the people calling. Wouldn't it be nice if recorded audio message files from only that person could be forwarded to your email address? Pamela lets you set up this email notification in either of two ways: without forwarding the audio file, or with the audio file bundled with the notification.

Pamela is Web savvy

If you can forward email notifications and slip in a sound file, then what's to stop you from creating an alternate Skype ID, calling Pamela, and having the message sent to some Internet server where it can be blogged or podcasted? Rather than code some geeky stuff such as XML or RSS, you can get the same results with Pamela's built-in capabilities.

So don't worry about the geeky stuff. Pamela has specific features that make blogging and *PamCasting* (Pamela's version of podcasting) easy.

Pamela is very punctual

Skype gives your customers, clients, and associates easy access to you from anywhere around the globe. However, *anywhere around the globe* can mean that people halfway around the world could try to reach you during the middle of their day — while it happens to be 1 a.m. your time. To help manage your presence and interaction with the broader global community, Pamela has the Time Scheduling function shown in Figure 9-17. In this way, Pamela keeps you in touch with others while also making sure that you get your rest.

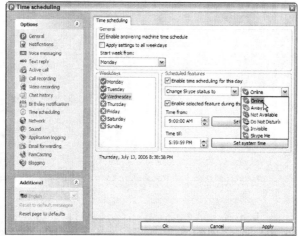

Figure 9-17: Pamela's Time Scheduling options.

Pamela is polite

It often happens: You are away from your computer and someone sends you a chat message. A polite reply to a single chatter is, "I'm away from my computer. My personal assistant, Pam, is responding with this automated message."

This type of response is good for letting single chatters know that you're not currently available. But in a multi-chat situation in which many people may be available to chat, this kind of automated reply can be downright rude. Pamela avoids the potential faux pas by discerning the difference between a single chat and a multi-person chat, and behaving the way you want in each situation.

Pamela is moody

As your list of Skype contacts grows, you may notice that mood messages for some of the more Skype-savvy contacts change regularly. You might see that some of these say, "I'm busy right now. Call me when I'm off the phone." A few minutes later, the mood message might read, "Okay, I'm available to take your call." Skypers can change their mood messages as they please, but you can have Pamela automatically change mood messages for you as you answer and hang up calls.

Chapter 10

Partying On — On the Conference Line!

. . . .

In This Chapter

▶ Starting and running a Skype conference

▶ Going beyond the basic conferencing facilities

▶ Moving files and sending links

▶ Viewing screens and using applications

. . . .

*T*alking one on one with your friend or colleague is a type of communication that Skype facilitates very effectively. You may also spend a lot of time in the company of others. When you're communicating in a group setting, you're probably accustomed to exchanging ideas, opening a dialog, debating, and even bantering. In short, you probably like to collaborate.

So, you're in luck! Skype conferencing makes collaborating with a group easy, and this chapter is your guide to doing so. In this chapter, we share some basic conference-calling techniques, show you how to use the conferencing capabilities that are native to Skype, and illustrate how third-party hardware or software can expand your conferencing capabilities. The combination of Skype and Web conferencing enables you to share files, show what's on your computer screen, allow others to remotely access other applications, send chat messages, and discuss ideas in a group setting. With these capabilities, you can do anything from running online classes to holding press conferences to strategizing a new product launch in your company.

The Benefits of Conference Calling

When a group of people gets together — with a specific purpose or function in mind — the group selects a corresponding venue to fit the occasion. Coming together in a Skype conference call is no different. The approach to conferencing with Skype is not a one-size-fits-all solution; instead, you can conference with others using Skype in three ways:

- ✔ **By calling multiple people at the same time:** This is the simplest form of conferencing with Skype — right out of the box. There's no additional software to download or special procedure to follow. All you're doing is connecting a group of people over Skype.

- ✔ **By adding Web conferencing and third-party tools to the mix:** For example, GoToMeeting and Unyte are third-party tools that extend your Skype-conferencing reach by handing over control of your computer screen and giving others on the conference access to your keyboard and mouse. Web conferencing transforms your conference calls into virtual meeting rooms. Quite literally, there is very little you can do in a face-to-face meeting that you can't do with Web conferencing and Skype together. If you've not investigated any tools like these, I predict that (as soon as you find out what you can do with them) you'll want to start using them . . . immediately . . . ASAP. Really, you won't want to wait.

- ✔ **By catering to a large crowd:** When you want to gather a large crowd — perhaps to hold a press conference, run a shareholder's meeting, or organize a relief effort in the aftermath of a hurricane — you can resort to facilities such as Skypecasting (see Chapter 14) and High Speed Conferencing (see Chapter 12). Skypecasting can accommodate up to 100 people at a time, and High Speed Conferencing can accommodate up to 500 people. Both of these services are free to Skype users.

The remarkable fact is that conferencing with Skype in all three venues has an astonishingly low barrier to entry, and the type of conferencing you can do for little or no cost is amazing.

What you can do with conference calls

The ability to make VoIP calls, be it PC to PC, or to a landline or mobile phone is great, but having conference-calling capabilities is really the icing on the cake. Conferencing opens up a world of possibilities. For example, suppose you bought a software product, only to find that it doesn't work properly. The software company claims that your computer is not configured correctly; the computer manufacturer claims the software company doesn't know what it's talking about. Everybody wants to point fingers, but nobody wants to take responsibility. No problem. Catch them off guard with a conference call and let 'em duke it out. We predict that your troubleshooting problem will be solved in no time at all!

Here are some other things you can do with conference calling:

- ✔ Plan a surprise party
- ✔ Have a telephone visit with someone in the hospital
- ✔ Hold a book club dialogue
- ✔ Podcast a panel discussion
- ✔ Conduct a language class
- ✔ Hold weekly status meetings
- ✔ Help real estate agents schedule a closing with all the lawyers, bankers, and other parties
- ✔ Connect doctors at multiple locations to confer about a patient

We encourage you to make your own list of ideas for conferencing via Skype. Just use it!

Applying good conferencing practices

A traditional way to conduct a conference is to bring participants together physically (say, in a conference room) and talk face to face. Years ago, that was the only way to hold a conference. A technology such as Skype changes the ground rules for conferencing, but with new capabilities and opportunities come new challenges.

Communicating by Skype conference call with several people at one time can quickly get complicated. For example, intended conference participants may be located in different time zones, or they may look at the scheduled conference call as a chance to run through a long list of unrelated topics. Here are some simple suggestions for applying good conferencing practices to your Skype conference calls:

- ✔ **Plan ahead for time-zone differences:** The people you want to bring together in a conference call can be located halfway around the globe from you or each other. A call you initiate at noon in Denver occurs at 3 a.m. in Hong Kong. Of course, if you regularly keep in touch with the various parties for your conference call, you can easily arrange to accommodate time-zone differences. You can imagine the potential for embarrassment when you schedule a conference call with people overseas and you call an hour later than planned because of a mix-up over Daylight Saving Time.

DST, GMT, UTC, and so on

Nope, the title of this sidebar doesn't contain acronyms for industrial compounds and chemicals or insecticides. GMT and UTC signify time zones (*Greenwich Mean Time* and *Coordinated Universal Time*, respectively), and DST stands for *Daylight Saving Time*, though just about everybody we know calls it Daylight "Savings" Time. Here are some fun facts about these time notations:

✔ **DST:** Benjamin Franklin published a proposal for Daylight Saving in the *Journal de Paris* in April 26, 1784. Actually, when ol' Ben proposed it, he suggested that people adjust their schedules and not their clocks. By Franklin's calculation, the good people of Paris could save 8 million pounds of candle wax each night if they moved their schedules around by an hour. Well over a century would pass by before the idea of Daylight Saving would take hold and we would adopt the whole ball of wax.

✔ **GMT:** Refers to the local time at the Royal Observatory in Greenwich, England. The reason people standardize on a time zone based in England is not to make telling time convenient for scientists local to that observatory or to cater to the elite. Instead, this location lies diametrically opposite the *International Date Line*, an imaginary line on the globe between one day of the week and the next. (You can travel a short distance to cross International Date Line and jump 24 hours ahead or behind, depending on the direction you're traveling. The International Date Line is positioned so that the date change occurs over the ocean, where it causes minimal disruption.)

Because of this position, the local time in the GMT time zone is high noon at the same instant the clock strikes midnight at the International Date Line.

✔ **UTC:** For practical purposes, the same as GMT, except that UTC time is synchronized to atomic clocks. Do you think it's a bit odd that UTC stands for Coordinated Universal Time? If you're into trivia, you can find lots of information on the Internet about why the acronym is UTC instead of CUT.

Here's the bottom line: Don't get confused if you see someone using GMT while others are using UTC. You can treat them as one and the same.

Avoid calling at the wrong time by chatting over Skype or sending an email in advance of the call. Use this message to confirm the call's date and time in each time zone — your local time zone and the local time zones of each participant — as well as the time in GMT (Greenwich Mean Time; see the sidebar "DST, GMT, UTC, and so on" for more on time zones). If you ask participants to check their local time against GMT time, you minimize the potential for missing conference calls because of the complexities of time zone differences.

If you use Outlook to schedule your meetings and have it integrated with Skylook (see Chapter 9) and your participants have Outlook, then the time zone differences are automatically reconciled in each person's Outlook schedule.

✔ **Establish a Plan B for your conference call:** The kind of people who regularly participate in conference calls tend to be busy people on the go. Their time is valuable, and in their hectic real world, not everything goes as planned. Suppose you lose your wireless or Internet connection right before the call is to occur. Or perhaps one of the participants has a last-minute change of schedule. Whatever the cause, you would do well to have a backup plan. It pays to set up an alternate date/time for the conference call. This is especially relevant if you are trying to pitch a new project to a prospective client, or some similar situation in which the people you're talking with don't know you very well or are not regularly in contact with you.

✔ **Avoid conference-call drift:** If you are holding a formal conference call on Skype, you'll want to keep the conference call focused. The simplest method for maintaining focus is to provide call participants with an agenda. Simply open a chat window to all the conference participants before beginning the audio conference; then, copy the agenda from your Word document and paste it into the chat window. Or even better than pasting text, paste a link to a Web page that holds the agenda into your chat.

What if someone takes the conference call via phone and has no Internet access? You can explore a multitude of services (many of which are free or at least have free trial periods) that allow you to send faxes from the Internet. If a participant has no fax machine available, just read the agenda at the start of the meeting and refer back to it as the meeting progresses.

Native Skype Conferencing

Skype conference calling doesn't require any additional software. If you have Skype on your Windows PC or Mac, the conferencing capability is already built in. The people you conference with don't have to use the same kind of computer platform as you do. They can use a wireless mobile Skype device, such as a Pocket PC or Windows Mobile, and because you can conference with anyone reachable using SkypeOut, participants can even take a conference call on their landlines and mobile phones.

Right out of the box, Skype allows you to connect groups of users, which is the essence of conference calling. Exactly how many people can you connect this way? Skype by itself allows five people to be speaking together at one time, but if you have enhanced hardware (such as a dual-core processor), Skype allows you to take advantage of this and double that number to 10. But effective use of Skype conferencing doesn't require you to rush out and buy a brand-new, top-of-the-line computer with a dual-core processor (it's a nice excuse, though).

A difference between Skype conference calling and the more traditional conference call systems is worth pointing out. With Skype, you conference call out to reach all the parties, rather than have them call in to you.

Starting a conference call

The mindset with Skype conference calling is simple and less structured than with traditional conference calling systems. When you set up a traditional conference call, you first reserve the availability of a dial-in number for a specific date and time, call or email each participant with the information, and then have everyone call the dial-in number at the scheduled time. But Skype's technology makes this level of prearranging unnecessary. Whenever you feel like it (and your intended participants are available), you can reach all the parties over Skype in one step and just start talking.

Follow these steps to making an easy Skype conference call:

1. **In your Skype client window, Ctrl+click the Skype contacts you want to include in your conference.**

   As you select multiple contacts, the green handset button you use to start Skype calls changes appearance to indicate that you're in conference mode (see Figure 10-1).

Figure 10-1: Click the Conference button, and the conference session begins.

2. **Click the Conference button, and Skype calls the selected contacts.**

That's all there is to it!

Skype gives you alternative ways to start a conference. You can start a conference call from the Skype menu by following these steps:

1. **Choose Tools⇨Create a Conference Call.**

   The Start a Conference Call window opens.

2. **Type the title in the Conference topic text box if you want something other than *Skype Rocks*.**

3. **Choose your conference participants.**

   You can select individual Skype contacts in the All Contacts pane and click the Add button (see Figure 10-2). Alternatively, you can click and drag a Skype contact into the Conference Participants panel.

Figure 10-2: Choose your participants and conference topic here.

4. **After you select the participants for your conference, click the OK button.**

As the conference organizer, you can use this same method (click a contact name in the All Contacts list and then click Add) to add people to an ongoing conference call. Of course, you can add callers only up to Skype's limit on the number of participants allowed.

You can include SkypeOut contacts, as well as Skype-equipped contacts, in your conference calls. Before making the call, just add a SkypeOut contact and supply a regular phone number in the contact information. Once that contact is added, you can immediately include the SkypeOut contact in any conference call you initiate. Of course, if there is a charge for the SkypeOut service (which depends on where you are calling), you will incur those regular charges.

Care to hold a conference call while you're on the go? Specifically, we mean when you're on the road moving at 55+ miles per hour. For a small monthly fee, your phone company may provide you with a device to connect your laptop to the Internet wherever you find a reasonably strong cell phone signal. This feature, along with Skype, gives you a roving conference center. Of course, you'll be conferencing from the back seat while someone else is driving. Safety comes first!

Reconnecting a dropped caller

Conference calling on Skype is a no-brainer. Not only is setting up a call super easy (see the preceding section), but staying connected is also quite convenient. During a conference call, connections can be dropped unexpectedly. For example, suppose that a conference call participant is talking on a cell phone, drives through a tunnel, and then loses the connection (in some jurisdictions, it is unlawful to drive and talk on a cell phone at the same time). When you use Skype, you can reestablish the dropped connection without having to restart the conference call, as follows:

1. **Click the Contacts tab to go to your Skype client's contact list.**

   In your contact list, all the call participants should be already selected.

2. **Ctrl+click the dropped contact and click the green call button on the Skype client window.**

   The dropped contact gets added back into the call, and you don't have to close the current connections.

Seeing who's doing the talking (Or what he or she is saying)

Now that you can conference call easily with Skype, you may get adventurous and hold your power meetings with participants overseas. A potential challenge is this: The participants you're speaking with may have foreign accents that make them all sound alike to your untrained ear and leave you scratching your head trying to figure out who said what.

A welcome feature in Skype lets you see who is talking. The feature shows up differently on the Windows and Mac platforms. On Windows, a glowing halo surrounds the icon of the person currently speaking. On a Mac, the sound meter level for the speaking participant increases. Figure 10-3 shows this feature on both platforms. Now you can really "see" who is doing the talking in your conference calls (it's seasonedskyper in both examples).

Macintosh Sound Meters *Windows Halos*

Figure 10-3:
Glowing
halos and
booming
sound
meters.

If your participants are chatterboxes, you may see more than one halo glowing at the same time. In this case, you needn't worry about who's saying what because you're unlikely to understand what they're saying, anyway.

Skype conference calls are handier when the participants can share a common chat window to type in text messages and provide live URL links during the conference call. Here is the easy way to start a *simul-chat:*

1. **In the Contacts tab of your Skype client, Ctrl+click the participants for your simul-chat and conference call.**

2. **Start a chat.**

   With your participants selected, right-click and select Start Chat from the resulting menu. A chat window opens. You may want to start the chat with a message like "I am about to start the conference call."

3. **Start the conference call.**

   On the Skype client window, your chart participants are selected. As more than one contact is selected, the large green button should display the conference call symbol. Simply click the button to start the conference call.

When you start a chat with your conference callers, you get these advantages:

✔ **You can keep a written record of the conference call:** The text chat can persist long after the conference call itself is concluded. Additionally, you can retain the text content of the chat as a summary and record of information exchanged during the conference call.

✔ **You can ask just one participant a question without bothering everyone:** If you need to, you can quietly conduct a separate *whisper chat* (where you can send a private comment to one of the participants without having all the others aware of the side discussion; see Chapter 6) while the conference is in session.

✔ **You can display discussion material:** It's easy to paste several paragraphs of text into a chat window or give your participants links to click while still carrying on your conference. You can even transfer files to the conference participants.

Transferring files while conferencing

While conference calling with Skype, you can transfer files to the connected participants. Obviously, you won't be able to transfer a file to any participant connected over SkypeOut. The last time we checked, conventional phones don't have the capability to send or receive files. Also, as of this writing, mobile devices running Skype (such as PDAs), don't have the capability to send and receive files over Skype. (But we won't be surprised if this feature is added in the future.)

When you're on a Skype conference call, you can transfer files to regular skypers by following these steps:

1. **Make sure that you have a chat window open to your conference participants.**

2. **Locate the file you want to send.**

3. **Drag the file into the chat window.**

   Dragging the file to the area you would regularly be typing in messages sends the file to all the people on the chat who are capable of receiving files. Users on mobile devices who can participate on chats may not be able to receive files.

 File transfers during a Skype conference call occur in the background while everyone's speaking in the foreground. Because keeping the high-grade quality of the voice line is important, the file transfer may run a little slowly. Slow transfer is noticeable if the file is several megabytes in size. You may be able to speed up the file transfer if you first place the conference on hold (by choosing Call⇨Hold⇨Conference Call) and then do the transfer. This technique of speeding up file transfers also works on Skype calls to individual Skype contacts. Simply place your party on hold before transferring the file.

Conferencing with Skype Plus Third-Party Web Conferencing Tools

Skype conference calling is convenient and practical. You can convene a group people to talk and exchange ideas, send paragraphs of text, and even transfer files. The richness of the conferencing experience and benefits get multiplied when you add Web conferencing into the mix.

Web conferencing in conjunction with Skype provides a number of advantages that you cannot get using Skype alone. You can:

- **Present a stunning visual presentation in a group setting**: Here's your chance to win over a group of people in a virtual boardroom setting. Not only can you run through your PowerPoint presentation, you can also "hand off" your presentation to another presenter midstream, showing what's on somebody else's computer. That's not all. You can create an interactive presentation that allows your conference participants to run and control the application on your screen from their computers.

- **Record your presentation or online session:** Web conferencing can provide for features such as creating an audio or audio plus video recording of your online meeting. You can use this recording for your own personal records, for rebroadcast over the Web, or for DVD distribution as an instructional video.

  When you record a conversation or meeting, be sure to adhere to any legal requirements in your jurisdiction about notifying parties that you are recording the online meeting.

- **Create the stunning visual presentation in collaboration with your colleagues:** Displaying your screen and applications in a collaboration mode is a very effective way to work. In this manner, you can build very effective presentations with the help of your colleagues but without requiring their physical presence.

- **Web conferencing tools can have some administrative aids:** If you plan to use Web conferencing and your line of work is professional services related, you may appreciate the fact that a number of Web conferencing services can log dates, times, and participants of online meetings. Capturing this information in the form of a spreadsheet, Web page, or text file may come in very handy.

When you tally the number of hours spent online, and the number of hours you save your clients (not to mention the reduction of travel expenses), you will find that dollar for dollar, next to Skype, which is effectively free, the added dollars you spend on Web conferencing is some of the best spent money in your budget.

Adding Web conferencing to Skype is easy to do. All you really need is Internet access (which you already have if you're using Skype) and to sign up with a service that provides Web conferencing. In this chapter, we describe two Web conferencing tools: GoToMeeting, a product of Citrix Online, and Unyte, a product of WebDialogs. There are plenty of others to pick from.

To help you understand how Skype and Web conferencing work together, we take a closer look at each of these Web conferencing products.

Web conferencing with GoToMeeting and Skype

Skype and GoToMeeting work very well together. To get started, you need the following:

- ✔ **Skype and an Internet connection:** You need a broadband Internet connection; a dial-up connection is not fast enough and cannot support both Skype and GoToMeeting at the same time.

- ✔ **A computer running Windows:** To host an online meeting with GoToMeeting, you must be using the Windows operating system. Intel-based Macs that run Windows also work.

- ✔ **GoToMeeting:** To get GoToMeeting, go to www.gotomeeting.com and set up a GoToMeeting account. You have these options for your GoToMeeting account:

  - • **Standard Account:** You can pay for the service a month at a time or sign up for a year for a discounted fee. A standard account allows you to have up to 10 participants + 1 for the host. You can explore the features of GoToMeeting by setting up a free trial account.

  - • **Corporate Account:** Gives you multiple user accounts but enlarges the number of participants to 25. Additional licensing options expand the number of participants to 200.

The features that we show in this chapter are based on using the standard edition of GoToMeeting.

There's a point worth mentioning about using Skype and GoToMeeting together. Depending on your Skype configuration, when you host a Skype conference call, you may have a limit of one host + four participants, or one host + nine participants. The standard version of GoToMeeting supports one host + ten participants. If you need to hold a combined Skype/GoToMeeting conference that exceeds a total of five people, you need to make sure that you are using the appropriate hardware such as a dual-core processor so that Skype can take advantage of it.

The number of users permitted on a Skype conference call is dependent only on the configuration of the computer used to host the call. You do not need to concern yourself with the kind of computers your conference participants use; only the one you are using to host the conference matters.

Here are some things you can do with GoToMeeting and Skype:

✔ **Hold a private online meeting:** A key advantage of using Skype in conjunction with GoToMeeting is that all the information exchanged can be 100 percent encrypted, including the information showing on the presenter's screen, the keyboard and mouse movements, chats, exchange of files, and the Skype audio conferencing. Because the Skype voice and chats are secure, you can even transmit the GoToMeeting password securely over Skype!

✔ **Hold an online interactive class:** Now you can run a virtual classroom in cyberspace and all your students can be anywhere in the world. You can teach them how to run a piece of software and give them turns at taking over the mouse and keyboard. You can give them a class assignment and later file-transfer the solution. Because you can switch presenters at any time, you can team-teach the class, and individual students can be made presenters. This is unlike anything the Show and Tell that we used to know in grade school!

For whatever it's worth, GoToMeeting gives you the ability to make audio and video recordings of your Web conference.

✔ **Solve a problem for a client:** If you happen to be a consultant and your client has a problem on one or more of their computers (it could, for instance, be a problem with some formulas in a spreadsheet), you have the ability to connect in over Skype and GoToMeeting and collaboratively work on the problem with the client. Skype makes it easy to transfer files, such as a spreadsheet, or software drivers, or whatever is needed at the moment. To clients, there are two other tangible benefits:

• **Security:** Clients may want to give you temporary access to their private data to fix a problem but not allow you to take custody over the data.

• **Software licensing:** Clients may be running some special software that has a per-seat software license fee or might be difficult to install on your computer. Accordingly, Web conferencing à la Skype lets you circumnavigate these issues altogether.

✔ **Team programming:** The idea is not so far-fetched. As you are developing some code for an application, you can use Skype and Web conferencing to develop an application jointly with a colleague who can be hundreds or thousands of miles away. Both of you can type and edit the programming code, and run the code on the spot. As you need to, you can transfer files back and forth using Skype. Having high-quality sound in Skype makes it seem as though your colleague is sitting in the room next to you. In this mode, you can work productively for hours at a time.

A chief benefit of this kind of team programming is the fact that you can improve the quality of your work product while shaving hours off of development, testing, and validation.

Setting up your GoToMeeting account

During the process of creating a GoToMeeting account, you provide some basic information including a your email address and password. After you are signed up, you can immediately start hosting online meetings. To do so, go to www.gotomeeting.com and click the button that says *Host a Meeting.* You are prompted for your email address and password, which is information you provided when you set up your account.

The first time you host a meeting, GoToMeeting automatically downloads and installs all the necessary files on your computer. Thereafter, you can host or schedule a meeting by right-clicking the GoToMeeting icon in your system tray and choosing either Meet Now or Schedule a Meeting.

Conducting meetings

After you establish an account and set up GoToMeeting, you can use it as much as you want. GoToMeeting puts no limitation or restriction on the amount of time you spend on meetings. GoToMeeting can also provide (at no charge) a dial-up phone number for the audio portion of your meeting. The dial-up number you are given is not a toll-free number. If you elect to use this service, all the participants who call would have to pay the toll charges. Fortunately, because you and all your participants can be using Skype, you can eliminate the need for a separate phone number and just use Skype instead. Setting up and using Skype as opposed to conventional phone service is simpler, works better, and has zero cost for Skype to Skype communications.

GoToMeeting works the following way:

- ✔ **GoToMeeting gives you a meeting ID:** When you set up a meeting, you receive a URL that you must disseminate to your participants. Although you can email this information, the easy way to invite participants is to send the link over a Skype chat. Participants join simply by clicking the link in their chat window.

  When the participants click the link, their Web browser opens a new Web page associated with the meeting. If they never used GoToMeeting before, a special Java applet is automatically downloaded and launched. It may be necessary for them to signify that they "trust" the applet and agree to let it download and launch. Participants are prompted to sign in and supply their name and email address. If you establish a password for your meeting, they are also prompted to supply that password. The whole process is automated and is a lot easier than it sounds.

- ✔ **Meetings can be prescheduled or on the spot:** You can set up online meetings that are scheduled in advance for a particular date and time, are recurring, or are impromptu. In a scheduled meeting, you can specify the meeting subject.

✔ **Meetings are kept private:** All the transmissions over GoToMeeting take place over encrypted channels. For extra security, you can require users to provide a meeting password. Additionally, if you use Skype for the audio portion of your meeting, for text messaging, and for file transfers, you can hold a complete online meeting that is 100 percent encrypted.

To start a meeting with Skype and GoToMeeting, follow these steps:

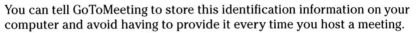

1. **Right-click the GoToMeeting icon in your system tray and choose Meet Now from the resulting menu.**

2. **Enter your email address and password when prompted.**

   You can tell GoToMeeting to store this identification information on your computer and avoid having to provide it every time you host a meeting.

   A GoToMeeting panel opens and displays a variety of controls for your meeting. You can optionally change the meeting subject by choosing File⇨Edit Meeting Subject in the GoToMeeting panel.

3. **If you haven't already done so, start your Skype chat and conference and get your participants online.**

   If you are unsure how to start a Skype conference, refer to the section "Starting a conference call," earlier in this chapter.

4. **In the Invite People (or Invite Others) section of the GoToMeeting panel, click the Copy/Paste tab and then click the Copy button.**

   The information about how to join the meeting is now in your clipboard.

5. **Click the input region of your Skype chat window (where you normally type messages) and press Ctrl+V to paste the meeting information for your participants.**

   Before pressing the Enter key, you can edit the meeting information. GoToMeeting automatically gives a dial-in number. Because you are already using Skype, you don't need to include this number, unless you want it as a backup to your Skype call.

As a host or meeting organizer, that's all you need to do.

Participating in meetings

If you are the recipient who is invited to a meeting over Skype, here is what happens:

1. **The host opens a Skype conference and chat with you and the other participants.**

2. **The host posts a link to the meeting on your chat window, and you click it.**

GoToMeeting's Mac support is limited

GoToMeeting version 3.0 and later has limited support for the Macintosh. Mac users who are running OS X 10.3.9 or later and who are not running Windows can join a hosted GoToMeeting session, but in a "read only" mode. Mac users can view a presenter's screen but cannot be given control of the presenter's mouse and keyboard.

Additionally, the Mac user cannot be made a presenter. If the Mac user is running Windows on one of the Intel-based Macs, then the user's Macintosh in Windows mode has all the privileges and features available to Windows users, including the ability to be given mouse and keyboard control, as well as being made a presenter.

When you click the link, your Web browser launches (if it is not already open) and goes to a page to see whether you already have the GoToMeeting Java applet on your computer. If you don't, it is downloaded for you and you are asked to trust it. Say Yes to trust it. You have the choice of selecting Trust Always or Trust for Just This Session. If you select Trust Always, you won't be asked this question every time you start a GoToMeeting session. Either choice is fine.

3. **When prompted, supply your name, email address, and, if necessary, password.**

   After the Java applet is downloaded, it prompts you for your name and email address. If the meeting host requires a password, you will have to supply this as well before you can enter the meeting.

When you enter the meeting, you may see a welcome screen or the presenter's computer screen (see Figure 10-4). The presenter is in control of the meeting. He or she can control what you are allowed to see. The presenter can choose to give you access to his or her mouse and keyboard. The conference organizer can change presenters at any time and can even make you the presenter.

Taking over the GoToMeeting reins

If the conference organizer transfers meeting control to you and makes you the presenter, a window pops up on your screen that announces as much. This window also asks that — before showing the contents of your screen — you close any confidential window and then click the Show My Screen button. Nobody sees what is on your screen until you say it is okay by clicking this Show My Screen button (see Figure 10-5).

If you need to opt out of being the presenter, simply click the Close (X) button in the top-right corner of the pop-up window.

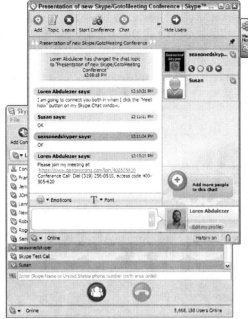

Figure 10-4:
GoToMeeting
conferencing
software
used in
conjunction
with Skype.

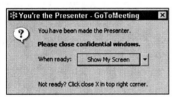

Figure 10-5:
When you
are made a
presenter,
nobody sees
your screen
until you
click Show
My Screen.

Whether you're the conference organizer or one of the participants, the GoToMeeting panel takes up a good deal of screen space. Click the "–>>>" icon at the top-left corner of the panel to collapse it.

"Unyting" Skype with visual communications

Unyte, appropriately named, brings together Skype and Web conferencing. The Unyte software integrates tightly with Skype and comes in two varieties, which you can download from www.unyte.net. Unyte Basic — the free version — allows you to display your screen to one other conference participant. The Unyte+ version allows you to share your screen and applications with multiple participants in a session. Unyte+ is fee based, and the fee depends on how many users you need to connect at one time. You can start with five users (which matches the low-end configuration of Skype) or, for a higher fee, you can have up to 25 users on a Web conference. Pricing is also based the duration of your subscription.

If you need to hold a Web conference that goes beyond the five-or-ten-participants limit of Skype, you will need an alternative source to handle the audio portion. The recommended source is High Speed Conferencing (www.highspeedconferencing.com), a free audio conferencing service for Skype users that can support conferences of up to 500 people. You can find out more about High Speed Conferencing in Chapter 12.

You can host an online meeting on Unyte or Unyte+ only on a Windows-based system. If you're on a Mac, you will need to be running Windows on Boot Camp or a virtualization software that allows you to run Windows.

When you host a Web conference on Unyte+, you have to specify a number of options:

- ✔ **What applications and desktop resources to share:** You have the ability to specify what participants will see when they connect to your computer (see Figure 10-6). For instance, you may be showing a PowerPoint presentation but, behind the scenes, running an Excel pricing spreadsheet that you don't want your viewers to see.

- ✔ **Allowing remote control:** When you make your desktop visible, you have the option to allow remote control from other computers. At any time during your online meeting, you can turn remote control on or off.

  When you allow remote control, all the participants in the online meeting can control your computer. Participants using Macs can remotely control your computer as easily as those using PCs.

- ✔ **Skype contacts and invitations:** Unyte gives you the option of selecting from your list of Skype contacts to invite to your online meeting (see Figure 10-7).

  After you invite the participants, a new Skype chat window opens with a message from you that says something like this:

  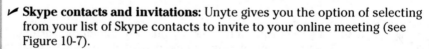

      I'm inviting you to view my desktop applications. Click this link to start:
      https://skype2.unyte.net/skype?s=268269141&k=29089&l=en-US

Web conferences are connected through Java applets running from participants' Web browsers, so it is not necessary for the invitees to be running Skype. It is not necessary for you to limit your invitees to only Skype contacts. You can also send invitations by email, copying and pasting the meeting URL into an instant message, or simply by reading instructions to other people over the phone.

When you invite non-Skype participants, you will have to resort to a separate audio system to accommodate your invitees. The one system that can handle this for both Skype and non-Skype users is High Speed Conferencing (see Chapter 12). This system is free to set up and use. Skype users get to call for free. Non-Skype users calling from regular landlines and cell phones incur whatever it normally costs to make a call to an assigned phone number for the call, which is based on the country the caller is dialing in from.

In addition to displaying screens and using applications, Unyte conference participants can draw or highlight relevant items on-screen, or use an arrow pointer to draw attention to important information (see Figure 10-8).

Figure 10-7:
Inviting
participants
from your
Skype
contacts.

Figure 10-8:
Ongoing
conference
session with
Unyte.

Chapter 11

Spicing Things Up with Great Gadgets and Add-Ons

Skype comes in four flavors: Windows, Macintosh, Linux, and Mobile. After you choose a flavor (or two), you can serve up Skype in any number of ways. Instead of a cup or a cone, you use devices such as laptops, desktops, wireless phones, mobile computers, or memory sticks to hold your chosen flavor(s). Then you can go for the toppings! Why not try adding microphones, webcams, wired or wireless headsets, or Bluetooth transmitters to your Skype sundae? You can also add software programs that record conversations, follow you with video as you move, or transform your image into a talking dinosaur.

With so many flavors and toppings (that is, platforms and add-ons) to choose from, we suggest matching your selections with how you use Skype for work or play. Here are some ideas:

✔ **For the frequent traveler:** If you need to take Skype with you wherever you go, try Skype on a laptop (or other mobile device) with a headset.

✔ **For the collaborator:** Suppose you need to keep track of ideas shared on Skype conference calls; try using audio recording software and transcribing software to capture every idea perfectly.

✔ **For the integrator:** If you find it handy to connect calls from other people's computers, try loading Skype — with your call list, chat history, bookmarks, and so on — onto a USB drive.

✔ **For the too-busy-to-sit-down entrepreneur:** If you find it difficult to stay tethered to a computer, try a Bluetooth headset and speech commands to manage your Skype communication.

 ✔ **For the frequent business caller:** Suppose you could make all your land-line calls and Skype calls from the same device by just pressing a button. You might want to plug in a dual-duty phone that handles both kinds of calls seamlessly.

No matter which devices, operating systems, add-ons, or software packages you use, one element remains the same: How you bring all these tools together can help you connect with others, work more productively, save money, and just have fun.

Giving Skype a Mobile Platform

Your desktop or laptop computer isn't the only platform you can use to run Skype. You have other fun options — for example, Skype-enabled phones and PDAs — that free you to take Skype wherever you go. One of these mobile platforms, plus an Internet connection, is all you need to hook up with your Skype contacts.

Skyping from your thumbtop

The notion of tucking away an entire communication and productivity center on a stick is pretty outrageous. But it's also true. Devices such as the U3 Smart Drive from SanDisk (shown in Figure 11-1) let you take your data and your programs with you. You can explore the current variety of U3-compliant software at the U3 Software Download Center (`http://software.u3.com`), which features device-ready programs for download and purchase. In what seems to be the blink of an eye, computing environments have moved from desktops to laptops to palmtops, and now we're into *thumbtops*. We can't help imagining that using such devices gives us access to one big information network where — like the Borg of "Star Trek" — we tap into the global mind (without the sinister overtones, of course).

We tried the U3 Smart Drive, which comes preloaded with Skype. This small device, in addition to backing up files, allows you to run programs directly from the smart drive without having to touch files from the computer's hard disk. With Skype preloaded, all you do is plug the drive into a USB port on an Internet-connected computer, start Skype, and log on to your account.

Running Skype from a U3 Smart Drive or other removable drive offers you portability as well as the following:

 ✔ **An increased level of privacy:** All your Skype-related information — chat history, bookmarks, notifications, and SkypeOut/SkypeIn numbers — remains on your portable drive. When you remove your portable drive, you remove your information from the host computer.

✔ **Freedom from downloading and installing Skype:** You aren't in the uncomfortable position of asking to download software onto someone else's computer in order to run Skype. A technology has got to be good when it makes you a better guest!

✔ **Added portability:** Pocket your data along with your programs in one tiny space. Everything is ready for transmission over Skype, just when you need it.

✔ **Generate confidence:** Knowing that your programs and data are accessible in another complete package and location — that is, tucked away on your smart drive — reduces the techno-jitters before a presentation.

✔ **Backup your peace of mind:** Even if you bring a fully configured laptop along, having a smart drive waiting in the wings as a perfectly rehearsed understudy reduces the worry that your show won't go on.

The programs loaded onto a U3 Smart Drive have to be prepared to retain all the information that accompanies the software. If you feel squeezed for space and decide to upgrade your drive from a 1GB to a 2GB or 4GB unit, you can just drag the contents onto the new drive.

Figure 11-1:
Take Skype
with you on
a SanDisk
1G U3 Smart
Drive.

Photo courtesy of SanDisk Corporation

Turning a handheld computer into a Skype phone

When you use Skype on a Pocket PC device, you can call, chat, look up a profile, change your Web presence, make SkypeOut calls, receive Voicemail messages, and participate in a conference. During an active call, you can even see the photos or images that skypers uploaded to their profiles.

Never fear. . . USB is here!

A group of business associates — still jet-lagged from the five-hour plane ride — gather in a wood-paneled conference room. They plug the laptop into a projector ready to knock the socks off of their client with an eye-popping presentation. As the laptop boots up and the image sharpens on the screen, the first associate pales. The second grows paler still, and the third is downright ghostly.

The program needed to run the presentation is missing, and the crucial spreadsheet showing the latest quarterly numbers is nowhere to be found. The fourth associate calmly reaches into her pocket, pulls out a USB drive, and plugs it into the laptop. She fires up the program they need, which is fully functional from the USB drive. Then she starts Skype (also fully functional from her thumb drive), taps into the local Wi-Fi, and scrolls through her Contacts list. She fires up a call, transfers the crucial spreadsheet file through Skype, and has the presentation up and running just as the CEO enters.

The heroine of this story is using a new technology and a new approach to the information age mantra "Back up, back up, back up!" In this case, what you back up is not just data; it's the whole software environment and the peace of mind that goes with it. Keeping a loaded USB drive on hand is better than popping an antacid tablet when it comes to calming nerves before a meeting.

But here are a few things you can't do (well, not yet, anyway). You can't:

- ✔ Send or receive a file through Skype
- ✔ Hold a videoconference
- ✔ Initiate a voice conference call

If you don't have Skype on your Pocket PC, you can get it from `www.skype.com/download/skype/mobile`. The creators of Skype recommend that you download Activesync software to connect your mobile unit and laptop/desktop PC first, and then download and install Skype for Pocket PC.

 If you have an older version of Skype for Pocket PC on your mobile unit, uninstalling it before putting on a new version is a good idea. To remove software from your Pocket PC, choose Start⇨Settings⇨System⇨Remove Programs⇨ Skype Technologies S.A. Skype. Then choose Remove and you're ready to start over with a new installation.

The HP iPAQ hw6900 (shown in Figure 11-2) is a personal digital assistant, a Quad band radio, a GPS receiver, a camera, an MP3 player, a computer running Windows Mobile, a number cruncher, a streaming video player, an Internet radio, and a smart phone. But wait, there's more! It's also Skype ready, right out of the box.

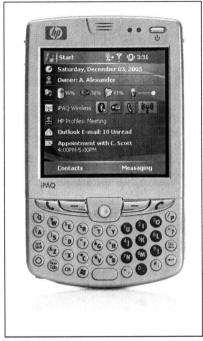

Photo courtesy of Hewlett Packard

Skype is already installed on the hw6900, so you won't have to figure out how to download a program as the first mobile thing you do. (You can save that tutorial for a Skype upgrade later.) However, a little fiddling is needed to connect to a Wi-Fi network. (*Wi-Fi* is the popular name describing the underlying technology of wireless local area networks [WLANs] based on the IEEE 802.11 specifications.) If you use a wireless computing device, you need a wireless network that's set up to access the Skype peer-to-peer network.

The iPAQ hw6900 rides on top of the Wi-Fi signal to provide Skype service. Your mobile unit will search for a signal; when you see your network name among the choices, press the network name with your stylus for one or two seconds until a context menu pops up, and then select Connect. If you don't see the Wi-Fi search screen, tap on Start➪Settings➪Connections➪Network Cards. Sometimes finding the wireless access is instant, and sometimes you do a considerable amount of tapping. But after you're connected, you can give yourself a wi-five and start skyping.

Before you can make the iPAQ hw6900 into a working cell phone, you must activate an account with a service provider and sign up for a fee-based plan. Then with Skype and Wi-Fi, you can make SkypeOut calls.

Skype looks almost the same (just smaller) on a mobile device as on a laptop. Figures 11-3 and 11-4 show various Skype screens on a Pocket PC platform. When you get to the program, you'll recognize all the icons. And you'll be skyping around town in no time if you follow these steps:

1. **Charge your iPAQ hw6900 (or other Pocket PC unit) and press the power button to turn it on.**

   The Today Screen — the little mobile desktop screen for those of us not in the Windows mobile universe — appears.

2. **Choose Start⇨Programs⇨Skype.**

3. **Tap the Tap Here to Sign In button on the Opening Screen (as shown at the top right in Figure 11-4) and choose an option on the sign-in screen, as follows:**

   • Create a new Skype account.

   • Log on to your own account.

4. **Enter your Skype ID (your Skype Name) and password where prompted and click OK.**

   Skype looks for a signal and logs in to your account (if you have a wireless network available). After you're connected, look through your Contacts list and check the Web status buttons that indicate your contacts' presence online. See Chapter 3 for information on verifying that a contact is on the Web.

5. **Select a contact from your list, tap the call or chat button, and skype away.**

Figure 11-3: How Skype looks when running on a Pocket PC (Part I).

Call icons | Contacts found | Message log | Msg templates

New Contact | Options Screens | Starting a chat | Voicemail

Tap twice on the screen for skyping options during an active call. Large call function icons appear on the screen and let you check a profile, mute your voice, add the caller to your contacts, start a chat, or hang up.

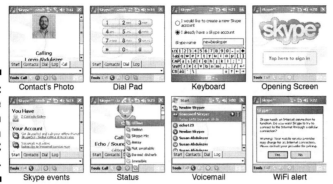

Figure 11-4:
How Skype looks when running on a Pocket PC (Part II).

| Contact's Photo | Dial Pad | Keyboard | Opening Screen |

| Skype events | Status | Voicemail | WiFi alert |

If you use the on-screen keyboard, it obscures the password field, so you may want to use the standard built-in QWERTY-style keyboard.

The iPAQ hw6900 comes with a clear cover that swivels up to reveal the touchscreen display. One of the authors (we won't say which one) spent an embarrassing amount of time trying to troubleshoot the unit, which wouldn't open any programs until we pointed out that the clear plastic cover was down, so that lovely touchscreen was not actually being touched. No amount of stylus tapping would start that puppy. After we figured out the lift-up-the-clear-cover hack — similar to the take-off-the-lens-cap hack — everything worked like a charm.

Skyping from your Wi-Fi phone: Netgear SPH 101 Wi-Fi

Those of you craving a truly portable Skype phone (but who don't need the turbocharged iPAQ) may want to look at the Netgear SPH101, shown in Figure 11-5. Weighing in at just a few ounces, this phone is extremely portable and fits comfortably inside your shirt pocket. Separately available is a cradle to keep the unit perpetually charged. Because of the phone's small size, the LCD panel is also small, but it is extremely crisp and the colors are bright, which makes the screen easy to read.

This nifty device taps into wireless networks and allows you to send and receive calls over Skype. Just imagine having no calling plans and no monthly fees. You simply find a public wireless network or connect into your own private wireless network and begin calling.

TIP

If you get a Wi-Fi phone, also get a SkypeIn number (with free Voicemail). This way, anyone can call you from a regular landline or cell phone as easily as people reach you on Skype from a PC.

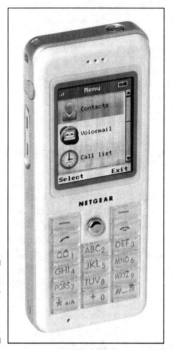

Figure 11-5:
The Netgear
SPH 101 Wi-
Fi phone.

Photo courtesy of Netgear, Inc.

"Wi-Fi"nding Your Way about Town

Do you have a Pocket PC but no wireless network? You can tap into many free or low-cost Wi-Fi resources as you skype on the go. Businesses, libraries, coffee houses, public parks, airports, train stations, hotels, and more have Internet access points. Skype has its own Wi-Fi service, SkypeZones, which offers 20,000 Wi-Fi spots you can tap into with a subscription. You can sign up for SkypeZones at `http://skypezones.boingo.com/search.html`.

Here are some other Web site locators to help you find a Wi-Fi spot anywhere you feel like skyping:

`http://anchorfree.com/`

`http://www.wi-fihotspotlist.com/`

Replacing Your Bulky Computer Phone

We've come full circle. Telephones started out bulky and wired; then they trimmed down and became wireless. Now we make calls over the Internet on computers, which are even bulkier and have more wires. Not to worry! More and more companies are producing Internet telephony devices that are smaller, wireless, and look just like . . . telephones!

Skyping around the house, and even in the yard

The Linksys CIT200 Internet Telephony Kit is a Skype phone for PCs (see Figure 11-6). The device consists of a handset with a dial pad and a color display that works just like a telephone. After you've synchronized your Skype information with the phone, you can scroll through your Contacts list on the display and press the button with the green telephone icon to dial out. There are no major lessons to learn, but there are major advantages to gain, as follows:

- ✔ **Finding available Skype contacts is quick and easy.** To the left of each contact name is a status icon that tells you who is online and who doesn't want to be disturbed. Push the toggle button up or down to scroll through your contacts. If you have a really long list, you can jump to a section quickly. Find the first letter of your contact's name on the dial pad. If you are looking for Loren, press the JKL button. Hold that key down for about two seconds, and the first name that starts with *J* is highlighted. Hold the key down again to jump to K names, and hold it a third time for L names.

- ✔ **Contacts' profiles are at your fingertips.** You can see some of a contact's profile information (other than a photo) if that contact has made it available. Pressing the button under the word DETAILS on your display shows a contact's language, age, and gender. Press the button under MORE to find telephone, fax, mobile phone, and home phone information. If your contact has not put this information into his or her profile, you won't see it in your Skype phone.

- ✔ **The Skype phone offers familiar phone operations.** To make a call, just select a contact and push the button with the green telephone on it. Push the red telephone button to end a call. You can put your caller on hold by pressing the green call button (you'll hear a dial tone). Press the Skype button at the bottom of the phone to see your contacts list. Choose a contact and press the button below the word DETAILS. Then, when the phone number information appears on the screen, press the same button (this time it will be below the word DIAL) and start another call. However, it won't be a conference call. If you're invited to a conference call, you can participate, but you can't start one.

✔ **Letting others know that you're online and available to call is a snap.** You can indicate your status to your Skype contacts through the Linksys handset by choosing Menu⟹OK⟹User Status to get the status choices. Select a status choice, and the rest of the world will know whether they can reach you on Skype.

Setting up the Linksys Skype phone is easy. But watch for the one hold-up: You can't plug it in right out of the box because it needs an overnight charge before the first use. After that, if the phone needs a charge, it will emit a little beep to let you know to put it back in its cradle.

Figure 11-6: The Linksys CIT200 Internet Telephony Kit.

Photo courtesy of Linksys, A Division of Cisco Systems, Inc.

Installing the phone's drivers

The Linksys CIT200 phone piggybacks on your computer's Internet connection and Skype program. A base station acts as the go-between. So, you need to install some software drivers to get all the pieces communicating. Linksys provides an install CD, so you don't have to search for the drivers over the Internet. However, you must install Skype and activate a Skype account prior to hooking up the Linksys phone. With that done, the installation process is simple:

1. **Plug the phone's base station into a USB port on your PC.**

2. **Put the Linksys phone in its charger cradle (after putting the two NiMH AAA batteries in the charger).**

   Make sure that you charge the unit for 14 hours before its first use.

3. **Insert the installation CD and run the setup.exe installer file (which should begin automatically).**

4. **Press the button on the base station (there is only one) to make sure that the base sees the phone handset.**

   If all is well, the handset will play a little music to let you know it's connected.

5. **Log on to your Skype account on your computer.**

6. **When the Skype API Security Screen appears, click the button that allows the Linksys CIT200 program to use Skype.**

Your phone is now skypified and ready for the acid test. Press the Skype button at the bottom of the phone, and if all goes well, your contacts will appear in the display.

Operating the phone: Details, details

The Linksys CIT200 phone is for Skype voice calling. You can't chat, start a conference, or send a file, but you can retrieve your Skype Voicemail if you have an account. This device has mute and speakerphone features, as well as a headset jack for earphones. You can connect up to four handsets to one base station, and if you have multiple computers in different rooms, you can connect one handset to four base stations. Choose a base station to connect to, and the Skype ID for that computer comes up on your phone. To connect to a base station:

1. **Select Menu (push the button below the word *Menu* in the LCD display).**

2. **Press the circular navigation button left or right to navigate to the SYSTEM menu.**

3. **Press the button under the word *OK*.**

4. **Press the circular navigation button up or down to select BASE SELECT. Click OK.**

5. **Navigate through the list of bases.**

   By default, they are named BASE 1, BASE 2, and so on. Click OK. Your phone displays the word SUCCESSFUL if you've made contact. You now have access to the Contacts list on that computer's Skype program.

Following are some points we found helpful to know when operating our Linksys Skype phone:

- ✔ **Press and hold 0 to get a + sign in your phone number:** When skyping out, your number, a plus sign (+), and the country code precedes the telephone number.

- ✔ **Keeping your Skype phone in range of your base station:** If you wander beyond the range of the Linksys Skype phone, it beeps in complaint. Simply move closer to the base station to pick up a signal.

- ✔ **Managing your audio devices:** The Linksys phone becomes the default audio device when you plug in the base station. Because your other audio does not play through the phone, you won't be able to hear any audio (other than a Skype call) unless you unplug it.

These phones are small enough to hide themselves under pillows and beneath papers. Ours buried itself under a mounting pile of chapter drafts. So we just pressed the magic pager button on the base station, and our Linksys Skype phone started singing up a storm. We followed the music right to the unit. (If only everything else we lose would do the same thing, we'd probably have a lot more free time.)

Getting the best of both worlds: Your phones are one!

The Philips VoIP321 Skype phone for Windows 2000 or XP actually does double duty. It is an Internet phone and a regular phone in the same handset (see Figure 11-7). A base station connects to both a phone jack and a USB port at the same time. You can switch back and forth between services by pressing a button: specifically, a Skype button for making Skype calls and a Call button to get a dial tone. Calling couldn't be more convenient.

The Philips VoIP321 has all the standard features of a wireless telephone, including a mute feature, speakerphone, and headset jack. When you use the Skype features, you can make voice calls, but you can't see video or pictures and you can't chat, transfer files, or send SMS (text) messages. Although the phone's display is not in color, the backlighting and large, clear letters make for easy reading. And you won't need a magnifying glass to read this phone panel. So whether you're using a landline or a Skype line, this phone gives you the same solid experience.

Installation for the Philips VoIP321 is a snap. The Setup Wizard does all the work and puts a shortcut icon on your desktop after installation. Here's just one hint: When the installer asks you to press HERE to set the audio preference, it literally means the word HERE on the installer screen, and not the Next button. The audio preference is set to recognize the phone by default.

Figure 11-7:
The Philips
VoIP321
Skype
Phone.

Photo courtesy of Philips Consumer Electronics

After installation, follow these instructions to activate the phone and make a call:

1. **Click the shortcut icon (which looks like a white handset on a blue background, with the words Philips USB along the edge) to activate the VoIP321, if it isn't already started.**

   A small icon appears on your desktop tray when VoIP321 is active, so you'll know whether you need to launch it.

2. **Start your Skype program.**

   The Skype Security API makes its appearance and asks whether it's okay to let the Philips Phone work with the Skype software.

3. **Select the radio button next to Allow This Program to Use Skype. Click OK if you plan to use the phone all the time.**

   Otherwise, you can decide every time you power up Skype by choosing the Allow This Application to Use Skype, But Ask Again in the Future option.

4. **To make a Skype call, press the easy-to-see red *S* button on the handset.**

   You see your contacts displayed along with their online status.

5. **Press the scroll key up or down to a contact's name and press the Call button (the one with the little green telephone on it) or the red Skype button; either one will connect you to your buddy.**

   You can make a SkypeOut call as well, but you still have to put in the country code by pressing and holding 0 (zero) on the dial pad to get the plus sign (+) before the country code. Then press the red Skype button to call.

6. **To make a telephone call the old-fashioned way, dial the number, press the Call button (the button with a green handset on it), and talk.**

   On this phone, you can also press the Call button, and then dial, and then talk. You're not penalized for getting the steps mixed up.

Basic calling seems pretty straightforward, but the following operations make using this phone system a bit more interesting:

✔ **Call waiting:** Suppose you are on a Skype call and someone tries to reach you on your regular line. Not a problem. A special call-waiting tone is played. When you hear it, you can end the Skype call and still take the landline call.

✔ **Conference calling:** You can use your VoIP321 with your Skype service to conference with two other skypers. Dialing one skyper, press the Call button (which becomes a Hold button when you have a Skype call in progress), dial the other skyper, and then press the Conference button to join the three callers. This experience is more of a traditional telephone conference, but within the Skype service.

✔ **Multiple handsets:** Because the Philips VoIP321 works with up to four handsets, you can have multiple Skype calls going at the same time on different handsets. But you can't have multiple handsets making different calls on the same landline simultaneously. The handsets are treated as extensions by the landline and as separate phones by Skype.

Getting Clearer Communications

You can operate Skype just fine from your desktop computer with its built-in speakers and microphone. But you may want to improve on the sound quality, ensure privacy, or otherwise free yourself from the basic setup's limitations. Headsets (both wired and wireless) can help you keep the noise down and bolster privacy, whereas adding a speakerphone can produce a clearer voice transmission on both ends.

And don't worry that adding extra gadgets will tie you down to your computer. Fortunately, you don't have to give up the habit of wandering from desk to refrigerator and back again while on a Skype call. You can simply clip on a wireless Bluetooth headset, voice activate your calls, or plug in a noise cancellation speaker phone while you're getting yet another piece of that coconut cream pie necessary to fuel all those Skype conversations.

Okay, we've made the process of adding accessories to your Skype operation seem effortless, and in many ways, it is. But some USB (Universal Serial Bus) devices should not be connected during installation until the software install program directs you to plug them into the computer. Make sure that you check installation instructions before you begin. The device you're hooking up may not work properly if you don't complete the steps in the correct order.

We'd also like to point out the following helpful tips for avoiding potential problems when hooking up headsets or other accessories:

- **Install device drivers *before* plugging in the USB device:** After device drivers are installed, plug in your USB headset first, before you launch Skype. Otherwise, your audio input and output may not come through the headphones properly.

- **On Apple Macintosh, use a USB headset/microphone:** Apple computers have mini-plug headphone jacks, but external audio input is either digital or USB based, so the USB headset/microphone combinations work best. All Macintosh computers have built-in microphones, so you can get away with using the computer mic and your iPod earplugs in a pinch.

Hearing better: Ready, headset, go!

Headsets help keep your conversations private because only the headset wearer (you) can hear what the caller is saying. And having a headset with a mute button adds an additional measure of privacy because you can push it to speak freely with someone in the same room without your Skype partner hearing what you say.

Voice echoing is another potential problem that can occur if one party doesn't use a headset while on Skype. Echoes happen when the computer's built-in microphone picks up the voice coming out of the computer's speaker. The caller hears his or her own voice projected back into the computer microphone. It's hard to talk when you keep hearing yourself repeating something a second later. Headsets and microphones help eliminate this feedback loop by putting a little barrier between the microphone and speaker.

Some headsets to consider adding to your Skype setup include the following:

✔ **The Premium Headset 350 from Logitech (see Figure 11-8):** This headset connects to your computer through a USB port. Besides volume and mute controls, the headset features noise cancellation and a microphone that pivots out of the way if you just want to listen to music or play a computer game.

✔ **The Internet Chat Headset (see Figure 11-9):** Also from Logitech, the Internet Chat Headset is among the variety of headsets that use miniplug input into the microphone and headphone jacks in the computer. Some of these models also have volume control.

✔ **The Lightweight Computer Headset Model 33-1187 from RadioShack (see Figure 11-10):** This headset doesn't have any bells and whistles. It uses mini-plugs and is an inexpensive but dependable choice.

Figure 11-8:
The Premium Headset 350 uses your computer's USB connection.

Photo courtesy of Logitech

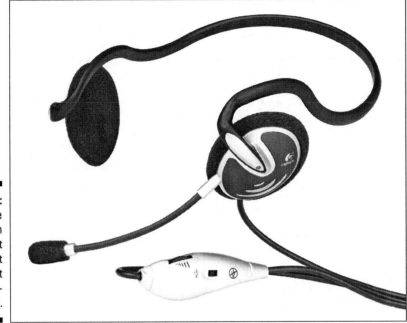

Photo courtesy of Logitech

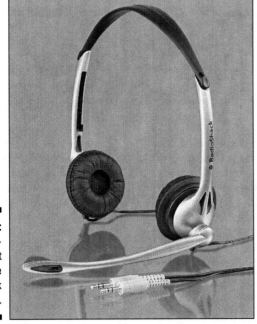

Figure 11-10:
An inexpen-
sive but
dependable
RadioShack
headset.

Photo courtesy of RadioShack Corporation

Saved by the headset

Imagine this scenario: Jack plugs in his headphones into the computer's USB port, positions them over his ears, and takes a deep breath. He launches Skype, selects Jill from the Skype Contacts list, and clicks the call button. His heart skips a beat as he hears her voice, a voice so soft that he raises the volume to maximum on his headset soft volume control. Nervously, he asks her to the prom.

Just as she starts to speak, Jack's Mom calls loudly for Jack to help set the table for dinner.

Jack blushes and quickly hits the mute button on his headset. Now he can hear Jill's reply without letting Jill hear all that noise. After all, asking someone to the prom is tough enough without being embarrassed by Mom's yelling in the background.

In this case, adding new technology to a Skype operation can be kind to nervous teenage boys summoning all their courage to ask a girl out.

Improving voice quality with a speakerphone

If you're on the go, wedded to your laptop, and wedded to Skype, then you're bound to get hooked on the Polycom Communicator C100S speakerphone for Skype, as shown in Figure 11-11. The most common first comments we hear from people who use this device is "If I closed my eyes, I would swear you're in the room standing next to me."

This is a speakerphone that you don't have to hover over, and you don't have to worry about a hiss or crackle during the conversation. You can carry on a Skype conversation easily from across the living room. The person you are speaking to would admit that the volume drops a little but that he or she still can hear you very clearly.

Many of us have endured speakerphones that handle noise cancellation by silencing the person who is not doing the talking and thereby eliminating the echo feedback looping from the speakers to the microphone. This noise cancellation method prevents two people from talking at the same time, which can prove frustrating to a caller trying to get a word in edgewise when the other person launches into motor-mouth mode.

The Polycom speakerphone handles echo cancellation flawlessly and lets both parties talk at the same time without being cancelled out.

Photo courtesy of Polycom, Inc.

Figure 11-11:
Improve
voice clarity
with the
Polycom
Communic-
ator C100S.

The C100S is simple to operate. After the initial software installation, just plug it into the USB port. The Communicator should automatically become the default audio device (mic and speaker) for your system and it works with just five buttons:

- **The Pick-up/Hang-up button:** Pressing the Pick-up/Hang-up button (on the right) lets you answer an incoming Skype call or hang-up a call in progress. Additionally, you can use this button to launch a call with whichever Skype buddy you have highlighted on your list.

- **The large center microphone mute button:** This appears as a micro-phone with a slash running through it; see Figure 11-11. Pressing this button lights up a bright-red LED to let you (and anyone glancing at the unit) know that you can speak privately.

- **The two volume buttons at the top and bottom:** Look for buttons with speaker icons in loud and soft mode; see Figure 11-11). These buttons on the speakerphone panel allow you to adjust the speaker volume up or down while keeping the microphone volume constant.

- **The blue Skype button (the *S* on the left):** This button enables you to launch Skype with just a push when the phone is plugged into a com-puter running Windows XP. If Skype is already running, pushing the button brings the Skype window to the foreground.

At the time of this writing, no software is available for the Mac OS X (although Polycom plans to add Mac driver support in the very near future).

Absent Mac OS X drivers, if you plug the device in without loading any software or drivers, the Mac operating system does recognize the speakerphone as a standard USB audio device (mics and speaker). Therefore, the volume controls are functional, as are the mute button and the red LED indicator light. If you have an Intel-Mac with Boot Camp (which allows you to run Windows XP) installed, the speakerphone behaves exactly as it does on any Windows-based PC.

The unit has built-in left and right microphones. Whether the family is gathering on Sunday to sing "Happy Birthday" to your grandfather a thousand miles away, or you're joined by half a dozen business associates spread around a conference table, no one will feel alienated from the conversation.

Everything about the Polycom speakerphone is well thought out, and here are some features worth noting:

✔ **The phone has a 3.5mm mini-plug jack:** This lets you carry on a private conversation using the ear buds from your iPod or any other stereo headphone without disturbing your associates in the office cubicles next to yours.

✔ **The speakerphone comes with a diagnostic software application:** This allows you to verify the operation of all buttons (plus the green and red LED lights), test the sound input from each of the left and right microphones, and generate a set of test sounds from the speaker.

✔ **The USB cable tucks neatly under the flip-open back panel used to angle the speakerphone:** A little opening in the side of the panel enables you to close the panel and lay the unit flat on a table top, without pinching the USB cable as it is connected to your computer.

Adding Bluetooth for Wireless Connections

You may love the privacy and clarity that headsets provide, but you may also hate that feeling of confinement you get from their wired connection. One way to cut the cord from the computer to your ear is to use a Bluetooth-enabled headset. *Bluetooth* is a specification (IEEE 802.15.1) for wireless personal area networks using an unlicensed radio frequency. (See the sidebar "What is Bluetooth and how did it get that odd name?" for just a little more info on this technology.) Bluetooth-enabled headsets look something like the odd implement jutting out of Lieutenant Uhura's ear on the original "Star Trek" series. In fact, the name *Uhura* means *freedom* in Swahili, so it's clear that she took her wireless communication seriously!

Look Ma, no wires

Bluetooth allows devices to communicate over a short range (usually easily covering the important desk-to-refrigerator distance) of about 30 feet. Some computers have Bluetooth built in, but you can add this feature externally if your computer doesn't. When you add it externally, you go through some software setup and a little bit of configuring to get the Bluetooth gadgets to link together. But after they do, it's wire-free sailing.

Motorola produces a number of wireless devices built into glasses, hats, jackets, helmets, and headsets. The device Motorola manufactures to use with Skype is the Wireless Internet Calling Kit Headset and Bluetooth USB PC Adapter, shown in Figure 11-12. The headset is a little over-the-ear clip-on unit for use with a PC. The clip part swivels out so that you can easily put it on and tuck the speaker part in close against your ear canal. The more complicated part is installing and syncing the Bluetooth USB Adapter.

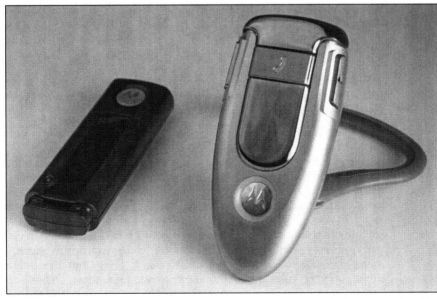

Figure 11-12:
The
Bluetooth-
enabled
Motorola
Wireless
Headset
H500.

Photo courtesy of Motorola, Inc.

To assist you with loading drivers for Bluetooth communication, Motorola provides one of those mini-CDs with its tiny headset (seems to be a theme here). If you're afraid to load this little CD in your computer, you can make a duplicate on a full-sized CD, and it works perfectly. We tend to do this (make a full-sized duplicate) when we're loading software onto an Intel-based Mac running Windows. We're afraid we'll lose the little disks into the slot-loading CD player.

Follow the steps to Bluetooth liberation

When you install the drivers from the Motorola CD, the Initial Bluetooth Configuration Wizard walks you through the process, but you still have to make some entries and choices:

1. **When prompted, select Headset from the list of devices you want to connect wirelessly to your computer and click Next.**

   If you use your wireless headset continuously, make sure that you check the box to start it automatically. Otherwise, you'll have to go through a synchronization procedure.

2. **In the same dialog box, choose to have a secure connection and click OK.**

   The Bluetooth activates and searches for a connected device. If you are using the Motorola headset, an icon with the model number appears in a list of connectable devices.

3. **Choose your headset from the list of devices and click Next.**

4. **When prompted, enter the code that comes with your device in the PIN Code text box to pair up the computer and headset (as shown in Figure 11-13) and click Initiate Pairing.**

   This is a crucial step to activate the headset-to-computer connection.

After you complete the installation of the Motorola H500, you need to press a button on the unit and hold for a few seconds to have it connect to the Bluetooth Adapter. A lovely circular blue light on the headset shines continuously after the computer discovers the device and the two shake hands.

While the connection is active, the Bluetooth earpiece enables you to hear any sounds coming from your computer: all the alerts, any music you play, and, of course, your Skype conversations. A teensy, but effective, microphone picks up your voice and minimal background noise, which is a very nice feature if your office, like ours, is in the pathway of four hospital emergency rooms, a police academy, a fire station, and rush-hour traffic.

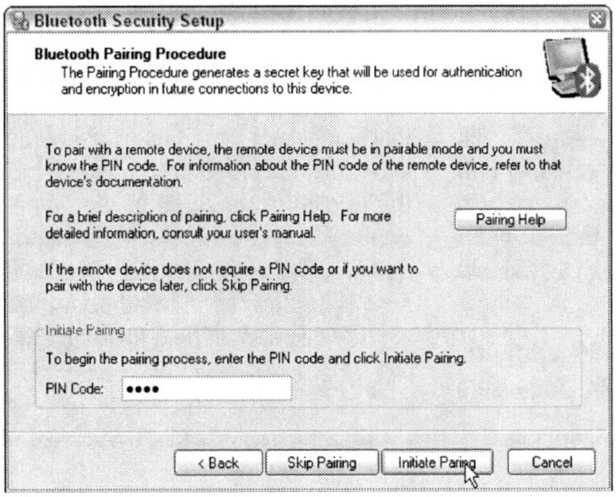

Figure 11-13:
Enter the
PIN Code
for pairing
Bluetooth
connection.

Bluetooth "Mac"inations!

If you want a wireless, over-the-ear headset that is fit for double duty — good for both Macintosh computers and cellular phones — you might try the Plantronics Explorer 320 Bluetooth Headset. This unit is designed for use with cellular phones equipped with Bluetooth, but it is surprisingly easy to connect to a Macintosh if you want to make hands-free Skype calls. Of course, your Macintosh must come with Bluetooth installed because this wireless device does not come with a USB transmitter (and doesn't need one). Bluetooth is a fairly standard feature on current Macintosh models, so if you have recently made a purchase, you should be good to go!

To connect the Plantronics Explorer 320 to a Mac, follow these steps:

1. **Click the blue apple in the upper-left corner of your desktop.**

   Choose System Preferences from the drop-down list.

2. **Double-click the Bluetooth icon.**

   The icon is blue and has a white, geometric "B" figure. If you don't know what the Bluetooth icon looks like, look at the geometric icon appearing near the top-right corner of Figure 11-13.

3. **Take your well-charged Plantronics 320 and press the power button for six to ten seconds (there is only one button).**

   The headset will start flashing red and blue.

4. **Choose 320 Plantronics from the list of devices; then, click the Set Up New Device Button.**

   The Bluetooth Setup Assistant opens.

What is Bluetooth and how did it get that odd name?

Bluetooth refers to short-range (about 30 feet) wireless technology. You can wirelessly connect a headset, mouse, and keyboard to a computer, link two laptops to each other, or pair a PDA (personal digital assistant) with a computer and other PDAs.

Bluetooth technology was developed by the Ericsson organization in the 1990s. Because of this company's Scandinavian heritage, the name *Bluetooth* is particularly significant. Bluetooth is the English translation of the name of a tenth century Danish King, Harald Blatand, known for unifying Denmark, Norway, and Sweden. So, if King Bluetooth connected countries, connecting devices should be a piece of cake.

5. **Click Continue.**

6. **From the list provided, select the radio button next to the word *Headset* and click Continue (again).**

   The Setup Assistant will search for your Headset.

7. **When the device name appears in the window, click Continue (yes, again).**

   On the next screen, the computer makes contact and gathers information about your headset.

8. **When the computer finishes, the Continue button becomes active; click Continue.**

9. **Enter a passkey to pair with your headset as prompted and click Continue.**

   The passkey information comes with each headset's documentation. The computer contacts your headset, and you are ready to make your wireless calls.

10. **Click Quit to close the Bluetooth Setup Assistant.**

Expanding Your Options with Software Add-Ons

Skype comes with many options, but you can greatly expand its native functions by using additional software. You may already have some software that you can use in conjunction with Skype. For example, Macintosh computers come with speech recognition software that you can train to work with Skype commands. And some programs that record audio for podcasting can also record your Skype conversations and voice messages.

Your voice is my command! Getting your Macintosh computer to listen and obey

Macintosh computers have built-in speech-recognition software installed on every machine. *Speech recognition* is yet another liberating technology that frees you from your keyboard by replacing keystrokes with vocal commands. When you turn on this feature, you have access to a long list of *speakable items* (words that your computer recognizes as input), including commands that let you open and quit applications, request the date and time information, switch applications, and shut down your machine. You can make some commands application specific, and you can make others work no matter what program you're running.

Activating your Mac's speech recognition

You can pair the Macintosh speech recognition feature with Skype by downloading and installing a set of commands *(Speakables for Skype)* that are scripted just for Skype. But before you install Speakables for Skype, or any other speech-recognition scripts, you must activate the speech-recognition feature on your Mac. To do so, follow these steps:

1. **Select the blue apple icon in the upperleft corner of your screen. From the drop-down list, choose System Preferences.**

   The System Preferences window appears. This window displays all the system preferences as icons. Notice the microphone icon labeled Speech.

2. **Click the Speech icon (the microphone).**

   The Speech dialog box pops up.

3. **Select On to activate Speakable Items, as shown in Figure 11-14.**

4. **With your microphone of choice showing in the Microphone drop-down menu, click the Calibrate button.**

5. **Calibrate your voice by speaking the provided phrases into the chosen microphone when prompted.**

   Each phrase, when recognized, will flicker in acknowledgment. You can adjust a volume slider if the computer is not recognizing your speech.

6. **Select an option next to Listening Method.**

   You can activate speech recognition continuously or when a key is pressed. Letting the computer listen to everything — especially in a noisy environment — is probably not a good idea. You never know what sound will trigger a command, so we recommend that you keep the default setting where the computer listens only while the Esc key is pressed.

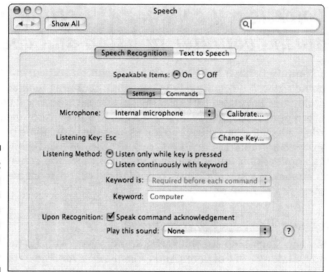

Figure 11-14:
Activating
Speakable
Items in the
Macintosh
system
preferences.

After speech recognition is up and running, you can test it. Press the Esc key and say one of the built-in commands out loud. For example, try "What time is it?", "Tell me a joke," or "Open iTunes." If you're not sure of what commands are available, you can:

✔ **Click the arrow at the bottom of the floating microphone icon that appears when you activate speech recognition:** The floating icon is round and displays a microphone image, the name of the key you have to push to activate Speakable Items, and an arrow for additional speech commands.

✔ **Find more commands:** Open the Speech preferences, choose Commands, and then click the Open Speakable Items Folder button.

✔ **Find all existing items the computer should recognize:** If speech recognition is working properly, just press the Esc key and say, "Show me what to say." This command brings up the window with all the existing items that the computer should recognize, and you don't have to hunt for the right folder with the right commands.

✔ **Try some of the listed items:** After you've found the Speakable Items list in the Speech folder, try some of the listed items.

Sometimes you need to say items a couple of times to get the recognition started, but after it works, using voice commands can be addictive. Watching someone tell a computer what to do is rather baffling to casual observers, but if you're used to walking down a crowded street speaking into an ear-mounted cell phone, those strange looks shouldn't bother you a bit.

Adding Skype's speakable items

With speech recognition active, tested, and working, you are ready to download and install Speakables for Skype scripts. These scripts are available at www.speakables.com/. Follow the directions on this Web site to download and then follow these steps to make the connection to Skype:

1. **Click the Speakables for Skype shortcut icon installed on your desktop.**

   A Speakables for Skype dialog box containing a table for adding contacts and commands opens, as shown in Figure 11-15, and Skype launches.

2. **Click OK in the security screen that pops up to tell you that another application wants to use Skype.**

3. **In the Speakables for Skype main menu, choose File⇨Refresh Contacts and then click Save to save your changes.**

   All your contacts are now imported into a drop-down list in the Speakables for Skype program.

4. **Select a contact and type a voice command.**

   In the Speakables for Skype window, note that you have two columns: Skype Contact on the left and Commands on the right. Select a contact from the Skype Contact list on the left and, under Command Phrase on the right, enter the voice command you want to say when you make a Skype call (see Figure 11-15).

   Your calling command can be anything from the mundane "Call Vincent" to the edgier "Yo! Vinnie."

5. **Press Enter when your command is finished; then, start telling Skype what to do.**

   After all, it's nice to know someone is listening to you.

Figure 11-15: Adding custom commands to Speakables for Skype.

You can make any phrase a speakable item in a snap. Activate speech recognition on your Mac by pressing the Esc key (after turning on speech recognition in your System Preferences). Then, select a file or document name, or even a URL, and say, "Make this speakable." A dialog box appears, inviting you to add a command.

Try this:

1. **Highlight the URL** `www.skype.com`.

2. **Press the Esc key and say, "Make this item speakable."**

   A dialog box appears, saying, `Go to "www.skype.com" When I say` (fill in the speech prompt in the text field).

3. **Enter the word or phrase you want to say to go to this Web site in the text box provided.**

   For example, enter the word *skype.*

4. **Press the Esc key to activate voice recognition.**

5. **Hold down the Esc key and say "skype."**

   Your browser opens up to that Web site.

Congratulations, you've just added another voice command!

Bluetooth headsets do *not* work with speakable items on the Macintosh. The alternative wireless solution is a Plantronics CS50-USB headset, which uses a different wireless technology and has a range of up to 200 feet.

Speakables for Skype has recently released a version of its software for Windows. Instructions and a free demo are available at `www.speakables.com/Skype_Windows.htm`.

Recording your hits (or just a few memory joggers)

You never know when you may replay a conversation in your mind — trying to remember all the details you forgot to write down, all the jokes you wish you could repeat, and all the family stories from your grandmother you want to capture. Instead of relying solely on your memory, you can find a number of ways to record Skype conversations using software, hardware, and a combination of both. We've included a small sampling of the possibilities here, but the Web is filled with tempting demos that are worth exploring.

For example, we suggest checking out the following Web sites for starters:

```
www.hotrecorder.com
www.podproducer.net/en/index.html
www.applian.com/freecorder/index.php
www.scodor.com
```

To get the latest scoop on recording applications, Skype has its own section for current programs (both tried and untried) at:

```
http://share.skype.com/directory/messaging/
```

If you're lucky enough to have a robust Skype Contacts list filled with fascinating, informative, clever, and scintillating conversationalists, you may want to install voice-recording software to preserve the witty repartee. Of course, remember that you should record conversations *only with permission* from your contacts. Recording conversations without all parties knowing may be illegal and is certainly impolite.

The following list outlines a few voice-recording products:

✔ **Call Recorder:** This product — an inexpensive add-on you can download from Ecamm Network at `www.ecamm.com/mac/callrecorder` — is designed for the Macintosh platform and makes your Skype calls into QuickTime movies. After you install Call Recorder, a Recording dialog box (shown in Figure 11-16) shows up right in the Skype application, so you don't have to open a second program. Call Recorder is simple to operate and gives you handy options:

- You can choose to record automatically, which is nice if you tend to forget to activate the record button, or you can start the session manually. You can have the recording controls visible all the time so that capturing important conversations on the fly is easy. Also, the quality of the recording can be set at high, medium, or low.

- Call Recorder creates QuickTime files and records separate audio tracks for incoming and outgoing voices. If you want to work with this file in an audio-editing program, you have a little more control than with one-track recording solutions. If you have the QuickTime Pro software, you can quickly get each track individually from the movie file using the extract button in the movie properties window. You can also use QuickTime Pro to export the file to another format.

Figure 11-16:
Make your
Call
Recorder
settings for
Skype here.

✔ **HotRecorder for VoIP:** This recording software for Windows computers works with Skype as well as other applications to capture audio. You can find and purchase this program at www. hotrecorder.com. The files created by this program are .epl files, which is a proprietary format used only with HotRecorder. In order to replay those files in other programs, you must convert them. HotRecorder provides a utility that you can use to save these audio files as MP3s, OGG, or WAV files.

✔ **SAM (short for Skype Answering Machine):** This is a Skype recording plugin for the Windows platform available at http:// kishkish.com/sam. SAM not only records conversations but also stores and organizes the recordings by time or participant. You can configure SAM as a regular answering machine that greets your Skype callers in the same way that a landline answering machine does.

With SAM, all your messages are stored digitally with names, date, time, and length of call for each contact. You can respond to these messages with Skype's callback feature right from the Message pane. This program offers a cost-effective way to build in a simple message center for a small business.

For larger businesses in need of recording, tracking, voicemail, and conference calling and Skypecast recording, the two major programs of choice are Pamela and Skylook, both of which we cover in Chapter 9.

Podcasting your Skype personality to the world

Skype is a rich environment for producing podcast content. As a step up from casually recording a Skype conversation (to help recall meeting details), some ambitious skypers are producing programs for broadcast (for step-by-step instructions, see Chapter 14). You can create podcasts of interviews, poetry readings, performances, news reports, and conferences over Skype. *Skypecasts,* the large-scale audio conferences posted on the Skype Web site, can be captured, edited, and converted into podcasts as well (see Chapter 14 for much more about Skypecasts).

Software programs such as Call Recorder, HotRecorder, Audio Hijack, SAM, Pamela, and Skylook are all capable of providing audio content for a podcast. However, some veterans of the podosphere prefer more rigorous recording equipment, and podcast studios can get pretty geeky. Getting truly clean audio is a tricky business, especially when your environment is virtual.

Podcasting For Dummies (also from Wiley Publishing) is a good resource for information about scores of gadgets and techniques for producing podcasts. This book even has an overview of equipment and editing software that's useful for getting the best audio production from Skype.

Transcribing your Skype recordings

So, you're busy recording anyone who will let you capture a conversation and you're filling up folders with MP3s of interviews, meetings, stories, and plans on every subject. Some of those audio files may be a great source of information to have in written form. For example, much of the research we did while writing this book exists as audio files of recorded Skype calls. As much as we would like to play our audio recordings for you, the practical requirement is that these recordings become paper and ink. To make this transformation, we had to convert what we recorded to what you're reading.

Listening to hours of Skype recordings while typing out a transcript was a huge job made easier by the wonderful software program Transcriber, shown in Figure 11-17 and available at no cost from `http://trans.sourceforge. net/en/presentation.php`.

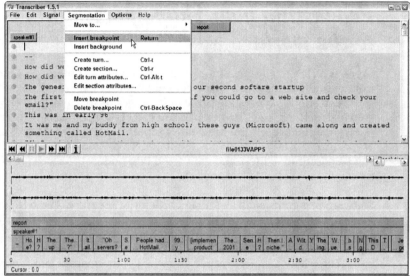

Figure 11-17:
The
Transcriber
program in
action.

This program's transcription features allow you to

- ✔ **Import MP3 files**: Think about all those Skype conversations you've been recording. You can export your transcriptions to files as text or HTML.

- ✔ **Create labels:** Labels can be based on the names of individual speakers. These names are incorporated into the timeline.

- ✔ **Create new sections:** You can segment audio portions and sync them with written files.

- ✔ **Zoom and telescope the timeline:** This feature enables you to jump to any part of a long recording very quickly and easily.

Transcriber shaves hours off the task of transcribing long audio sessions, organizing text by topic and speaker, and publishing the transcript in print or on the Web (for more information, see Chapter 14).

Video and animation add-ons

Third-party developers are tapping into Skype videoconferencing and forever changing the way we look at ourselves. Literally. Avatar animation products — for example, Talking Headz, Festoon, and Logitech's Quickcam Fusion — walk us through the process of becoming dinosaurs, pumped-up soccer players, femme fatales, emotional emoticons, and, our favorite, a hairy eyeball with an attitude. Each of these software packages helps you create your cartoon soul mate and show your silly side to the world.

Avatars are Web-based animated characters; virtual stand-ins for you, me, or anyone who wants to create an online personality ready to mix it up with other available avatars (see Chapter 5 for more about avatars). The idea behind using avatars with Skype is to play with video conversations, add character to instant messages, and just have fun in a new way on a new playground. These software packages work closely with Skype video messaging and launch at the same time you start up Skype.

The avatar animations are cleverly developed. When you speak or record a message, the avatar mouths your words quite convincingly. Some avatars respond to mouse tracking, with their eyes and heads following the cursor as it moves. Some have added personality routines, sighs, sneezes, and rolling eyes designed to elicit a response from a viewer. The following three avatar animation programs have unique offerings, so we recommend investigating each one:

- **Gizmoz Talking Headz** (www.gizmoz.com): Encourages you to build your avatar image from scratch and makes no pretense about keeping it serious. The characters, gags, jokes, sounds, and animations are indescribable, so visit this avatar farm and see for yourself.

- **Festoon** (www.festooninc.com): Provides for video conferencing with up to five people (a very practical side of this product) each transmitting from their own webcam. Festoon also provides video cutouts called *Eye Candy*, which are still pictures with gaps where the faces should be so that you can insert your own image.

- **Logitech QuickCam Fusion's Video Effects software** (www.logitech.com): Maps a still picture of your face to your chosen avatar to coordinate speech, or if you want, you can become a talking shark. As does Festoon, QuickCam Fusion sometimes uses webcams.

Guess Who's Calling? Customizing Caller ID

Just because you can talk to the whole world for free doesn't mean that you want to answer every Skype call. Sometimes you just don't want to talk to your lovable but endlessly chatty friend. So, a little call screening is in order.

Of course, Skype has a built-in caller ID. You can see the user name and picture (if one is posted), of each caller. But our eyes are not always on the computer screen, and it seems a little retro to have to run across the room to answer a Skype call just as we do for landlines (remember those?). Luckily, for every problem, no matter how small, there is an enterprising and fun solution.

Know who's calling (and no peeking!)

You can customize caller announcements with a software product plugin for Skype running on Windows called Wizztones (www.wizztones.com). With Wizztones, each contact can have a different ringtone (see Figure 11-18). You can keep any of the Skype ringtones, record your own, pick from a list, or even have a custom ringtone messages synthesized. To synthesize a custom ringtone, you upload a message (in text) and Wizztones will digitize it and add it to your messages list instantly, without leaving the Wizztones software interface! Wizztones also provides standard phrases (in addition to ringtones) to announce each caller — just in case you can't come up with a snappy phrase to announce that Uncle Ralph is calling!

If you want to use Wizztones to change a ringtone, follow these steps:

1. **Click the Wizztones icon (which looks like a blue planet with two big white Z's inside) in your Windows task menu bar.**

   The Wizztones options window opens (see Figure 11-18).

2. **Select a contact in the Skype contacts list.**

3. **Select the Other Available Ringtones option from the ringtones window.**

4. **Select the ringtone you want to associate with your chosen contact.**

   If the contact is your mom, you can select Your Mom Is Calling from the ringtone list. Of course, you can also choose The Boss Is Calling as her ringtone. (But beware, ringtones can be very revealing.)

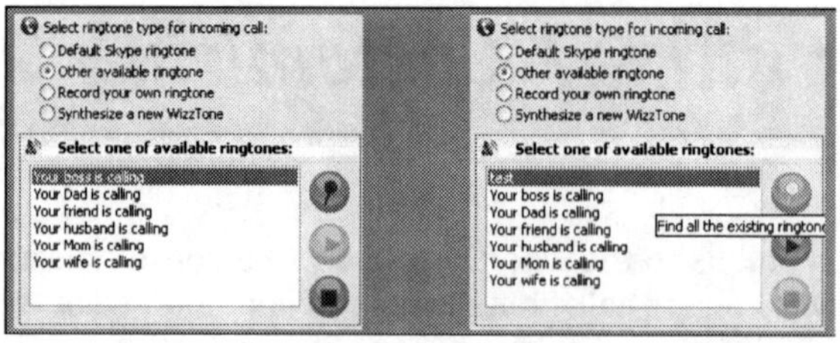

Figure 11-18:
Wizztones
ringtones
selection
window.

If you want to create your own ringtone phrases, you can do that, too. Select the Synthesize a New Wizztone option, enter the text to be synthesized in the text field when prompted, and click the little recording button to change the text to speech. Just sync the phrase with your contact, and when Mom calls, Skype announces, "Uh-oh, this is a call I can't refuse," or whatever phrase you choose to let you know what you're in for.

Creating your own caller announcements without downloading a thing!

Skyping on a Macintosh computer gives you a few cool built-in features for customizing sound input and output. One that works beautifully with Skype pulls in your contact user name information and transforms text to speech, with no typing on your part. Just accessing Skype preferences taps into this Mac utility, and you can use it to create personalized caller announcements (see Figure 11-19).

Getting Skype to announce a caller with a unique phrase of your choosing is simple. All you need to do is follow these steps:

1. **Launch Skype.**

2. **In the Skype menu, choose Skype➪Preferences➪Notifications.**

3. **From the Events pop-up list, choose Incoming Call.**

4. **Click to activate the Speak Text field.**

    The default phrase "You have an incoming call from "@"!" appears in the window. There is no need to enter a name. When you get a call from the chosen contact, somehow, magically, the "@" symbol is replaced with your contact's name.

5. **If you want, type over the text in the Speak Text field (just be careful not to type over or replace the "@"! portion of the text).**

    So, instead of hearing only "You have an incoming call from Loren," you can add to it, making it "You have an incoming call from Loren. That's the best news I've had all day!"

When you edit this incoming caller message text, the announcement changes for all your contacts, so it's wise not to be too specific when crafting the perfect announcement.

Figure 11-19:
Customizing
Skype
announce-
ments on a
Macintosh
computer in
the Events
window.

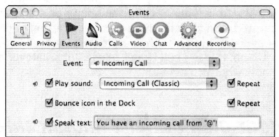

You can add this feature to all your contacts at the same time. In a busy office, or when you are facing a deadline, you may want to be notified imme- diately if your client is available, is calling, or is sending you a file. The basic ringtone only gives you the information that someone wants your attention. When you hear a name, you have a heads-up on what to say. These kinds of personalized announcements can skyrocket your productivity.

The following announcements are among those that can be customized in Macintosh events:

- ✔ Your contact "@" is available.
- ✔ You have an incoming call from "@"!
- ✔ Message from "@".
- ✔ You have an instant message from "@"!
- ✔ You have received an authorization request from "@"!
- ✔ You have received a list of contacts from "@"!
- ✔ You have received a file from "@"!
- ✔ File transfer with "@" is completed.
- ✔ File transfer with "@" has failed.

Finding codes with the long-distance widget

Widgets are handy Macintosh computer shortcuts to scores of applications. Click the widget icon and a widget dashboard appears. You can add, move, rearrange, and download widgets of all sorts. Calendar widgets, stock market widgets, webcam widgets, newsfeed widgets — there's a widget for every inclination, personality, and taste. So, appropriately for a Skype book, there is a Skype Widget, as shown in Figure 11-20.

The Skype Widget has a field to put in a SkypeOut number. If you are not sure of how to dial a long-distance number, the widget is very helpful. Say that you want to call Brazil. You know the main number but not the country code. The Skype Widget lets you pick a country from a list and then provides the right symbol and code for that call. After you put in the country, the calling rate appears to let you know how much this call costs. Click the phone to dial the number and activate Skype. Your Skype software opens and places the call.

Figure 11-20:
The Skype Widget for Macintosh OSX computers.

Shopping at the One-Stop Skype Shop

Combing the Internet for gadgets is an awful lot of fun. But if you want an efficient way to deck out your personal Skype Center, you can log on to Skype.com and click the Shop button. An online catalogue features phones, webcams, headsets, speakerphones, and other Skype accessories. Skype tests the products listed on the Web site and certifies that they work properly with Skype. Those products that are officially approved have a Skype Certified seal of approval.

Skype also posts a developers forum, `http://share.skype.com/directory`, where you can find loads of accessories, software, services, video and animation add-ons, enterprise software, messaging solutions, plugins, and even an entire section on everything for Skype from Japan. Many of the software and plugin offerings are free downloads, and each product is followed by a lively forum of product reviews.

You'll find a wonderful sense of playfulness within this section of the Skype site. Most of the products are not officially Skype Certified, so you're on your own here, but these products extend Skype in some of the most unusual, adventurous, and productive ways. It's worth combing through the descriptions and reviews, and even to take a chance and download some plugins. There are some very cool surprises here.

We are on the constant lookout for new and interesting gadgets, doodads, plug-ins, and services you may want to explore. Have a look at our updates on our Web site at `www.skype4dummies.com`.

Alice in Skypeland

While writing this chapter, we three authors felt as though we'd fallen down the rabbit hole and landed in the secret world of Skype gadgets. The boxes full of wondrous goodies just kept on coming — piling up in the corner of the living room, spilling over with wires, manuals, batteries, cradles, shipping peanuts, and excitement. Every box was an adventure waiting to happen.

We opened each one, rolled up our sleeves, furrowed our brows, plugged, configured, wired, synced, called, pinged, paged, launched, saved, sweated, and laughed. As we took yet another item off our leaning Tower of Skype toys, we called vendors (over Skype, of course) to ask endless questions and received endlessly patient answers. Our laptops lay on the breakfast bar, the dining room table, the couch, and the window seat as we moved from product to product. We groaned when we were baffled and whooped when we figured something out.

The living room has not quite recovered from the onslaught of gadgets, and we can truly say, neither have we. So we'll stay down here in the rabbit hole a little while longer. We're just having too much fun.

Part IV
The Professional Skyper

The 5th Wave   By Rich Tennant

"For 30 years I've put a hat and coat on to make sales calls and I'm not changing now just because I'm doing it on the Web in my living room."

In this part . . .

*T*ime to put on your gray-flannel suit and well-shined shoes for some formal skyping. Or better yet, forget all that and skype in your pajamas — but just remember to get dressed before that videoconference! This part points you toward some of the myriad ways in which you can put Skype to work for your business, saving money while increasing productivity. We show you how Skype can, for example, help you hold down travel costs, make customer service easier and more responsive, and hold videoconferences with one or even 500 people — the latter at no extra cost beyond your Internet service.

But don't think that by "professional" skyper we mean only work and no play. This part also describes what a rich resource Skype is for personal development and socializing. See how to tap into Skype communities all over the world for love, friendship, and education on every topic imaginable. Also, find out how to start your own Skype community with a Skypecast.

Chapter 12

"Skypifying" Your Business

Keeping a business alive and thriving is no easy task. Aside from managing people, you have to control costs, think fast, and be equipped to outperform your competitors. So what does this have to do with Skype, and why might you want to consider "skypifying" your business? Although we cover the features mentioned in this chapter in more depth elsewhere in the book, we offer this chapter to concentrate on ways that Skype can assist you in your day-to-day business tasks. Using Skype in your business amounts to much more than just calling a person from one PC to another over the Internet.

Also, if you do decide to seriously entertain the thought of "skypifying" your business, you need to be aware of the security concerns (unfounded!) associated with Skype. This chapter also includes some general information about how you can allay these fears from an IT (Information Technology) perspective.

Skype in Your Business

If an important part of your business is communicating with other people, whether for purchasing, connecting to your customers, or other ways, the chances are pretty good that someone in your company is already using Skype. According to the creators of Skype, roughly 30 percent of all Skype users regularly use Skype in their business. No matter how you slice it, this amounts to millions of people having adopted Skype as a standard business tool. It behooves you to know a little more about what Skype can do for your business, as well as how to control any risk or downside associated with adopting the technology.

Be aware, however, that Skype is not intended to be a replacement for your phone system. Here are some reasons:

- ✔ **No emergency call capabilities:** Skype cannot be used to call emergency numbers such as 911.

- ✔ **SkypeOut and SkypeIn calls don't carry the same high fidelity experienced with Skype-to-Skype calls:** Although the sound quality of both SkypeOut and SkypeIn keeps getting better with each new version of Skype, these services still have a ways to go before catching up with regular Skype calls.

- ✔ **Microphone-headsets or speakerphones are needed with Skype:** Not everybody is comfortable wearing headsets or using speakerphones. There are dualphones that can send and receive regular phone calls and Skype calls. They work well, but not everybody knows they exist.

- ✔ **Answering an incoming call on Skype requires that a device be turned on:** Many people are in the habit of shutting off their computer at the close of a business day. This means that Voicemail needs to be enabled. The good news is that when you sign up for SkypeIn, you automatically get Skype Voicemail for free.

The passage of time and new Skype and VoIP technology are sure to overcome these limitations.

Meanwhile, however, because you can't replace your entire phone system with Skype, why bother adopting it? Here are some reasons to consider:

- ✔ **It's not an either/or proposition:** Skype enriches the communication process by allowing you to do things not generally available on a telephone. You can use Mood Messages (see Chapter 3). You can transfer files (as explained in Chapter 6). You can combine audio conferencing and text chats. You have video. All this points to a feature-rich system that is remarkably inexpensive.

- ✔ **Skype can save your business money:** Many businesses that are service oriented, such as consulting, can eliminate large portions of travel expenses through the use of Skype and Web conferencing. The benefit can amount to saving thousands of dollars per employee who travels.

- ✔ **Thirty percent of all Skype users regularly use Skype in their business:** You don't want to be left out in the cold when other businesses are using an effectively free technology to get ahead.

Meeting Core Business Needs with Skype

All businesses have similar, although not identical, core needs. Among these core needs are sharing documents, preserving privacy, controlling costs, and marketing.

Sharing documents

Everybody is used to moving files around. We do it all the time as attachments to emails. Whether sending files by email, file transfer protocol (FTP), or some alternative means, the process is sometimes fraught with obstacles: file size limitations, encrypted or executable files, and so on. Skype offers a quick and expedient way to leap-frog over these challenges when you have to get things done quickly.

Three aspects in particular are worth noting about file transfer on Skype:

✔ **Ease and convenience:** From an open chat window (see Chapter 6), you can just drag one or more files from your desktop and drop them onto the chat window.

✔ **Security:** With regular email, unless you specifically go through some setup in advance, the files you transfer and receive over email are unencrypted. Many popular email systems won't even let you send encrypted ZIP files. Skype allows you by default to send files securely.

✔ **No predetermined file size limit when transferring files:** Although Skype is not a speed demon with file transfer, it doesn't complain about file size. By comparison, many business email servers complain if attachments are more than a couple of megabytes and prevent such large files from being transferred.

Skype enables you to carry on a conversation while you are transferring files over Skype. Keep in mind that you have multiple services (voice and video versus file transfer) competing for a limited resource (your network bandwidth). Skype audio trumps the Skype file transfer service. To speed the file transfer, try hanging up temporarily so that your voice and video transmissions don't compete with file transfer. You don't need to close a Skype chat, though — it's not a resource-intensive service.

Preserving privacy

In a world in which hackers and competitors would love to unscrupulously exploit your communications, protecting your business is imperative. With Skype, you can conduct the audio and video portions of a Skype conversation, as well as text chats and file transfers, as encrypted sessions. This means that when you are carrying on a conversation in the middle of, say, Chicago's O'Hare International Airport on a conference call, your competitors can't pry into your information no matter how hard they sniff the packets. They can't get into your voice communications, file transfers, or chats. With regular phones, however — both wireless and landline — privacy is not even an option!

If you are using SkypeOut, whether for a one-to-one conversation or on a Skype conference call, or someone uses SkypeIn to reach you, the portion of the communication going to and from the telephone is *not* encrypted.

A little padlock icon appears on the lower left portion of your Skype window during the call, signifying that your call is being securely transmitted.

Managing costs

These days, who isn't on a budget and trying to shave off a few dollars? At the same time, it would be nice to expand your business communications options. Because just about everybody is connected to the Internet, you can use Skype in addition to your phone system as a communications platform for a no-cost or low-cost solution.

To help you manage costs, Skype offers a service called Skype Business Groups, which allows you to purchase Skype credit in bulk and apportion it to members of your business or even selected customers. Rather than tie up large amounts of capital for specific resources for specific people, you can allocate it where you need it, when you need it. To get started with it, go to the Web site at `https://secure.skype.com/groups` and follow these steps.

1. **Sign in using your Skype Name and password.**

   A Control Panel – Business Development Web page appears that allows you to create groups, add users to those groups, and purchase Skype credit for those groups.

2. **Click the <u>Create New Sub-Group</u> link.**

   On the page that appears, enter a descriptive name for your group, such as Sales Team, in the text input field and click the Create button.

3. **Click the <u>Add Users</u> link.**

   A page opens on which you can add users to any of your groups.

4. **Enter the Skype Names you want to add to a group and choose from any of the groups you have created; then, click the Verify and Invite Skype User Names button.**

   The Skype User Adding Results page appears and you receive a notice such as the following: "You successfully invited *Skype Name* to your Skype

Groups. *Skype Name* will shortly receive an email containing the invitation details. They will need to accept the invitation by logging into My Account (link provided in email) and accept or reject the invitation there."

Of course, the Skype Name that appears in the message corresponds to the one you enter. The invitees receive an email message inviting them to join a Skype Group. To join, they follow the link on the indicated Web page in the email, supplying their Skype Name and password to log in.

5. **Purchase Skype Credit for the Group.**

You have the option to add Skype credit manually for a group or use what Skype calls the Auto Top-Up feature, which prevents Skype credit for the members of your group from dipping below a dollar balance that you set.

The principal advantage of Skype Groups is that you can centralize and more easily manage the purchase and use of Skype credits on an organization-wide basis.

You don't need to restrict your purchase of Skype credits to SkypeOut, SkypeIn, and Skype Voicemail. Other services are available to your Skype Group. For example, Skype offers a translation service on call 24 hours a day so that you can conduct your business with your overseas business partners, and you need Skype credit to use this service. If you have a team of people negotiating a contract or supporting business operations, it's more cost efficient make the translation service available to everyone on your team than to buy it separately for each individual.

To find our more about how Skype can benefit your business, go to `www.skype.com/business` and `http://share.skype.com/sites/business`.

Marketing

One of the strongest revenue drivers in marketing is to be in touch with your customers, both to speak to and listen to your target audience. Whether your goal is social networking, broadcasting a message, or contacting individual customers, Skype can help you to achieve your marketing objectives. The following sections describe some possibilities. (Also check out Chapter 16 for ideas about using Skype to promote your business.)

Hold a no-cost community gathering online

Reach a large audience by holding a Skypecast or conducting High Speed Conferencing. (See Chapter 14 for more about Skypecasts; High Speed Conferencing is described later in this chapter.)

Here's an example of what you can do through a Skypecast or High Speed Conferencing. Pharmaceutical companies and health care providers often engage in programs such as Direct to Consumer communications. These organizations can use online audio conferencing as a catalyst for the public to seek more in-depth information, including active dialogue with health care providers about the benefits and risks of a prescription medicine or a new clinical trial.

Stay connected with your customers and business associates

Skype equips you with an electronic way to hand out a business card: the SkypeWeb button. This button creates a live link that you can post on a Web site or include with an email signature block. A unique aspect of this link is that it shows your "presence," the Skype word for availability — which you can tell Skype to change automatically. Imagine handing someone a business card that changes its information when you are on vacation, out to lunch, or in the office. SkypeWeb buttons are clever and useful marketing devices. They can also be picked up by search engines because they are written in HTML code, the language of the Internet.

The SkypeWeb button just looks like your Skype online status button as it would appear on anyone's Skype window who happens to have you listed as a contact (see Figure 12-1). Also, the SkypeWeb button can be viewed and used by anyone who has Web access and cares to reach you.

Figure 12-1:
A SkypeWeb button with a "Skype Me!" status.

All you need to do is copy and paste a small piece of HTML code to Web pages you post. After the page is placed on the Web and viewed by anyone, the button is rendered to display your availability — for example, "Call me!"; "Add me to Skype"; "Chat with me"; "View my profile"; "Leave me voicemail"; or "Send me a file." It's a little magical. Sit at your desk, log on to Skype, and anyone can cruise your Web site and catch you in the office.

To create your SkypeWeb button, go to `www.skype.com/share/buttons/`.

Have a global presence from your desktop

Using a Skype Mood Message (see Chapter 6), you can send out instant global alerts to your Skype contacts. Simply post a short message next to your name in your Skype window, and all your Skype contacts who have you on their Contacts list will automatically see your message displayed next to your name in their Skype Contacts window. When you change the message on your machine, it changes on theirs. Neat, huh?

Maintain customer access at your fingertips with Skype

Your Skype Contacts list is not an island. It is easily bridged with your contacts in Microsoft Outlook. To find, and skype, anyone in your Outlook contacts list, simply have both programs running at the same time. In Skype, choose View➪View Outlook Contacts. All your Outlook contacts are instantly accessible from within your Skype Contact window. You can perform a quick search by entering a portion of a person's name in the text entry area for entering a Skype Name or phone number. As you enter the name, your list of contacts is filtered to show only those names, including your Outlook contacts, that match. If your Outlook contacts contain phone numbers, you can easily SkypeOut to them. It is an extraordinary timesaver, organizer, and therefore potential money saver.

Support your customer base with Skype

Many companies are beginning to discover the advantage of providing help-center call-in services via Skype. USRobotics, a major manufacturer of Internet and IT hardware, has adopted this model by placing a Skype contact link on its Web site (see Figure 12-2).

Figure 12-2:
USRobotics provides a customer support link on its Home page.

You can take this use a step further. For an organization with a traditional call center, adding Skype provides another access channel of support available at low or no cost to the customer as well as the company. Operators standing by can receive calls either by phone or by Skype.

Because you need only one Skype Name, you can have multiple operators on separate computers. When a call comes in over Skype, all the computers will ring and any of the operators can answer. The remaining operators are free to pick up the next inbound call. You can have as many operators as you need answering incoming calls, all on the same Skype Name.

Offer payment services

Integration between Skype and PayPal is in the works and expected within months of this writing. Your company operators can then not only answer callers but also securely receive payments for product support.

Mega Conferencing at Warp Speed

Sometimes you need some extra-large communication muscle to reach a super-sized audience. High Speed Conferencing (www.highspeedconferencing.com) is an application created by a company called VAPPS that works with Skype to reach your mega audience. As many as 500 people can connect to a conference using Skype or a regular telephone. The cost to you to host the conference is zero, nada, zip. If any of your attendees decide to call in from a landline or cell phone, they pay whatever they would be paying for a call on their particular phone plan. There are no supplemental or hidden charges of any kind.

Want to hold an audio press conference, stockholder meeting, or a special event for several hundred people? From your Web browser, go to www.highspeedconferencing.com and then follow these steps:

1. **Select Create Account from the menu options.**

   A new page appears in which to enter information for creating your account.

2. **Enter your Skype Name and email address associated with your Skype account in the entry fields of your Web page; then, click the Register link.**

   You receive a dedicated conference "room" number — that is, a virtual meeting place in cyberspace. For your conference room, you also receive a pair of personal identification numbers, or PINs, to log in as

Conference Moderator or as a Participant. Fortunately, you don't have to remember these numbers; this information is emailed to you when your account setup is complete.

High Speed Conferencing creates Web buttons (see Figure 12-3) that are similar to SkypeWeb buttons (described in the previous section of this chapter).

3. **Enter the names and email addresses for each of your conference invitees (see Figure 12-4).**

   The lists are designed to be used on a recurring basis, such as for monthly sales status meetings, because the invitee list is retained for further use. There are third-party applications that integrate with the High Speed Conferencing service and can construct conference invitation lists on the fly. One such application is a Customer Relationship Management (CRM) application called Salesforce (find it at www. appexchange.com).

   When your invitees receive their invitations by email, the message contains the Web buttons that they can click to automatically join the appropriate conference (see Figure 12-5).

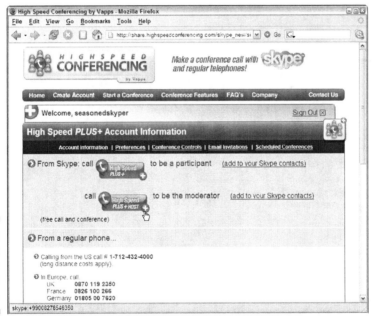

Figure 12-3:
Account
acceptance
with login
information.

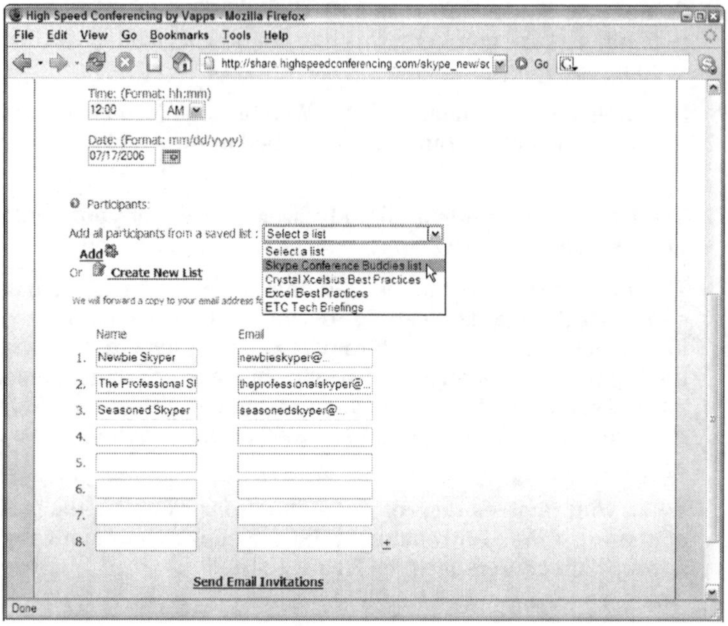

Figure 12-4:
Populate your invitee list.

Figure 12-5:
Email invitations with Web button links take you to your conference.

You can run your meeting using Skype for literally no cost with up to 500 participants.

Staying One Step Ahead by Being One Click Away

You and others in your business may spend a great deal of time performing searches on the Internet. When you find some key information such as a support hotline or technical support number, you want to pounce on it, call the contact, and save that number.

An assortment of Skype toolbars enable you to skype a person or contact directly from an application external to Skype, such as your Firefox browser or Office application, by directly clicking a phone number. You don't need to manually copy the number from the application and paste it into Skype to make a call. This feature means that your employees can work more productively and effectively, thereby reducing some of your operating costs.

Skype has specific toolbars for specific programs. These include separate downloadable toolbars for the Mozilla Firefox Web browser, Outlook, Outlook Express 6, Thunderbird (the Mozilla email program), and Microsoft Office.

The toolbar that Skype provides for the Firefox browser combines two facilities: a search panel to help you locate information with possible phone numbers, and a Highlight Numbers button that transforms phone numbers on a Web page into "skypeable" links (see Figure 12-6). To call, just click the green highlighted phone number or right-click the phone number and select one of the options on the drop-down list options for immediately starting a call, sending an SMS message, adding a number to your Contacts list, or just copying the number to your clipboard.

To download and set up the toolbar, go to www.skype.com and follow these steps:

1. **Click the Download button that appears near the top of the Home page.**

   The general Skype Download page appears with a list of downloadable items at the left.

2. **Click Skype Web Toolbar to display an expanded list of items immediately below it.**

   The expanded list contains the following:

   - About Skype Web Toolbar
   - Get It Now
   - Getting Help
   - Skype Web Toolbar Change Log

Search for numbers

Skype toolbar
enables/disables
highlighting numbers

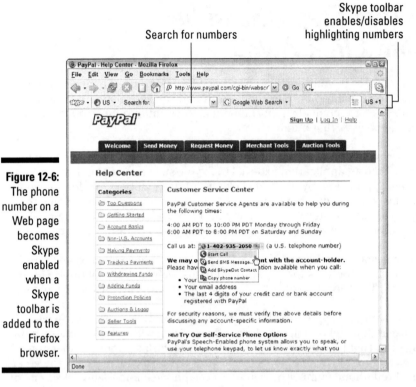

Figure 12-6:
The phone
number on a
Web page
becomes
Skype
enabled
when a
Skype
toolbar is
added to the
Firefox
browser.

3. **Click the <u>Get It Now</u> link and then click the Download button in the middle of the Web page to get the Skype Toolbar for Mozilla Firefox installer file.**

   Save the file to your hard drive.

4. **Double-click the file you just saved to open it.**

   If you have the Firefox browser running when you launch the installer, you are alerted that all Firefox windows will have to be closed before installation. When asked whether you want to close all Firefox windows now, click Yes.

5. **Choose the I Accept the Terms of the License Agreement option, click the Next button, and then click the Install button.**

   (If you want to change the default location for where you save the Skype Toolbar Firefox application on your hard drive, you must click the Options button instead of the Install button. Clicking the Options button enables you to change the save location before clicking Install.)

After you click the Install button, Firefox launches and a Software Installation window opens with the message A Web site is requesting permission to install the following item: "SkypeToolbarForFirefox."

6. **Click the Install Now button to install the Firefox extension.**

When the installation is complete, a window displaying your Firefox extensions is displayed along with the following notice: The Skype toolbar for Firefox will be installed when Firefox is restarted.

7. **Quit and restart Firefox to complete the installation.**

Phone numbers on your Web pages will now appear as buttons that you click to make SkypeOut calls (refer to Figure 12-6).

If your Skype program is not running when you click the button to call that number, it automatically launches. After you sign in, Skype immediately proceeds to dial the number as a SkypeOut call.

To fully understand the toolbar, you need to know the following:

✔ **Your Firefox browser needs to have the toolbar in an "on" mode (see Figure 12-7) to make phone numbers on the Web pages skypeable:** The toolbar acts as a toggle switch. Depending on its current state, clicking the toolbar icon once toggles it from an "on" state to an "off" state, or vice versa. When the toolbar icon is in the "on" position, phone numbers on all the pages you surf are immediately skypeable.

✔ **Overseas numbers sometimes appear without the country code, and you may need to set it:** The phone number provided on a Web page may contain a country code, but then again it may not. For example, a phone number may be listed by area code and phone number, without any country code. If the area code is 207, the phone number listed could be in the state of Maine, but it could just as easily be a phone number in London.

Unless you provide some information, there is no way Skype can know what country to dial when placing the call. If no country code prefix appears next to the area code and phone number, click the flag icon appearing to the immediate right of the skypeable link. Clicking this icon enables you to select a country (see Figure 12-8) and have the corresponding country code returned.

Figure 12-7:
Clicking the
toolbar icon
toggles its
on/off state.

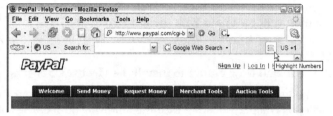

You can use the toolbar without actually making calls in Skype. Simply right-click the phone number and select Add as a SkypeOut Contact in the drop-down list. As you accumulate contacts, you can then use a program such as Skylook (see Chapter 9) to transfer your contacts to Microsoft Outlook.

You can get plenty of value from Skype without having to spend money on SkypeOut minutes. You can perform Internet searches on toll-free numbers (using SkypeOut to call a toll-free number is free). For example, you can create a search term such as `phone number IBM 1-800` to find a listing of various toll-free numbers at IBM.

Figure 12-8:
Switching
the country
context from
the U.S. to
the U.K.

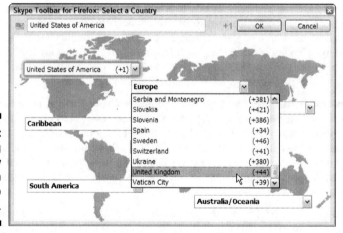

A variety of Internet search engines and Web sites enable you to locate toll-free numbers.

Addressing IT Security Concerns

Skype was originally designed to be a consumer-oriented product, but because of its secure transmission of audio, video, text, and files, it has quickly gained traction in the business world. These very benefits can also raise the hackles of Information Technology (IT) departments. To allay and address these concerns, Skype is incorporating features to make it IT friendly. Specifically, an IT department can deploy Skype in an organization while maintaining the ability to

- ✔ **Prohibit Skype altogether:** If you have a highly secure IT environment, it may be critical that absolutely no external communication with sources outside your network segment take place. In such cases, prohibiting the use of Skype in any form would be prudent.

- ✔ **Disable file transfer capability of Skype:** Skype file transfers are secure and encrypted. This means that a person who has access to the public Internet and to sensitive files can securely transfer those files to an external third party. The only way to control the situation is to disable the file transfer capability of Skype. You do so by making changes to the Windows registry of the computer running Skype and preventing unauthorized users from changing registry settings. This task is beyond the scope of this book.

- ✔ **Restrict Skype to route its content through a specific firewall or proxy server:** In a networked environment, the issue may not be so much the security of sensitive documents or files. It is easily possible for a person to securely upload a sensitive file to an external source from a Web browser. You don't need Skype for that. What you may need is the ability to scan an incoming file that is transferred from Skype and that may contain a virus signature. One way to assure the scanning is to route all Skype network traffic through a specific firewall or proxy server.

- ✔ **Disable all Skype APIs:** Even if Skype works in a manner that is acceptable to the practices and policies of your organization, you may be wary of other programs that can tap into the features of Skype through the Application Programming Interface, or API. In that case, you can disable the Skype API so that other programs cannot interface with Skype.

Restricting features within Skype is not simply a matter of selecting some item from a menu and voilà, it's done. Correctly configuring security entails much more than making changes to Skype itself. If you disable a feature such as the ability to transfer files, you also have to restrict the end user's access to administrative features of the computer. Otherwise, your end users can simply undo the settings.

In short, making changes to just Skype alone is not enough. You must secure the whole computing infrastructure of your organization.

Chapter 13

Exploring Skype Communities

. . . .

. . . .

Not even Gulliver in his travels could hold bragging rights over the variety of scenes, pursuits, opportunities, and enterprises populating the new Skype landscape. Every existing endeavor, as well as every endeavor made possible by the Skype peer-to-peer technology, appears redesigned as a Skype community. The speed at which new connections are made between people of like minds and interests is astonishing. As fast as you can read this chapter is as fast as Skype communities assemble. This chapter opens the door, just a crack, to an array of social, education, commerce, and special interest communities for you to explore.

The people we meet, do business with, socialize with, and learn with are no longer limited to random acts of luck (or misfortune). We can pick communities to join, leave communities we dislike, and start our own communities if there are none out there representing our obscure passion for playing the Hawaiian nose flute. In this chapter, you find out how to take a class or join a book club using Skype, where to find a Skype partner to learn a new language, how to use Skype to kindle new friendships (and romances), and what hot new Web communities are adopting Skype at their core so that their members can start to talk to one another.

Finding Your Perfect Skype Venue

As Skype becomes a more common way to connect, more Web communities are adding it to their menu of choices. The Skype contribution, instant voice messaging, adds immense value to existing tools such as emails, discussion threads, and text messaging. Listening and speaking are, after all, our most natural way to get to know each other. You can hear someone's personality in his or her voice as much as reveal your own. No wonder Skype is becoming a mainstay of some of the most active online social settings, such as the following:

✔ **Social forums:** Web-based search and communication tools designed to let you find other people with common interests for friendship, romance, conversation, or business.

✔ **Online classes:** Web-based groups led by instructors involving a formal sign-up process, or small study groups organized through search tools and having no formal teacher or registration.

✔ **Virtual conferences:** Web-based conferences for large groups (100 participants) led by one or more people on any topic imaginable.

Looking for love with a little help from Skype

Online dating is not new, but online dating with an instant, private way to speak with a potential date is a welcome addition to the experience. Skype changes the nature and experience of Web romance in some significant ways:

✔ **Talk to a prospective date:** Skype voice and chat tools let people hear each other before giving out phone numbers and addresses. The biggest benefit in using Skype to socialize is that you can create a direct and immediate contact with a potential friend.

✔ **Connect using cutting edge online tools:** Sites such as www. someonenew.com feature Skype as the primary incentive to use their portals (see Figure 13-1). Another online dating site, www.verbdate. com, advertises itself as the place "Where People Speak before They Meet." Unlike other sites built around those in the know about Skype, www.verbdate.com registration screens include a Skype flash tutorial, an explanation of the technology, and a link to www.skype.com to download the software.

✔ **Skype from within each online community:** On the site www. someonenew.com, the primary way you connect to another member is by clicking his or her Skype Name featured in an online profile. The link is active, causing Skype to open and begin a direct call to your potential new friend. If that "someone new" is online and willing to accept your call, just speak up!

✔ **Take the guesswork out of calling:** Making contact with a new friend, and possibly a new romantic friend, can be scary. When you call with cell phones or landlines, you have no idea whether anyone will pick up or how you will be received. Skype lets you — and your new dating partner — broadcast your availability to receive calls through online status icons, giving you a little time to build up your courage.

✔ **Connect privately and safely:** You don't need to give out a phone number in order to receive, or make, a direct call. You can skype from your computer to another computer or to a regular phone without revealing any personal information. Your Skype profile shows only the information that you choose to make public. So you can really begin to know someone before trading addresses and phone numbers, and you can do so without feeling uncomfortable about keeping that information secret.

Figure 13-1:
Skype is integrated into www. someone new.com.

Discovering Birds of a Feather Who Skype Together

Joining a social community has never been easier than in this age of virtual societies. This section introduces a couple of communities you may want to explore.

Beboppin' with Bebo

Bebo (www.bebo.com) is a Web site whose sole purpose is to create community and connect people. Its eye-popping profile pages, filled with the latest multimedia and interactive ingredients, let members show their personalities. After you see, hear, and interact with a Bebo profile page, you probably soon know whether you have found a friend to talk to over Skype. On their profile page, Bebo users can

- **Share playlists:** Bebo users can post favorite music clips on profile pages. These clips open in a colorful MP3 player, and you can listen to them on the spot. Click a link to automatically copy the playlist to your own Bebo profile if you like the selection.

- **Post a gallery of friends:** A section of your Bebo profile page displays Bebo friends you've added. Rather than a barebones text buddy list, you see all the photos your online friends uploaded to their own Bebo home page. Each photo has a link to each individual. You can easily begin to develop a large circle of friends with common interests. We searched for Bebo users who play the fiddle and use Skype. Bebo returned a long list, and each of those skyping fiddlers had galleries of friends. The potential for building a vibrant, personal community is built into the Bebo profile interface.

- **Grow a fan base for your favorite band:** Another section of the personal profile page is devoted to posting pictures of musical groups. Each picture is linked to a separate Bebo Band page listing information about the band, comments about the band members, tour dates, songs, albums, a gallery of fans, and the band's home page.

- **Show a movie:** Bebo allows you to upload a short movie clip onto your Bebo home page. Many Bebo movies are homemade; some are commercial clips. You can also click the <u>Add To My Homepage</u> link to put a Bebo movie you like onto your own profile page. Bebo's interactivity is all about sharing.

- **Draw a reply to a whiteboard message:** Bebo has a section on each profile page for drawing or writing on a blank whiteboard. It's an unusual area because each whiteboard drawing becomes a posted message to which other Bebo users can respond. In place of traditional conversation threads, whiteboards supply colorful, imaginative image threads. Mostly they are just another way to have fun at this online party.

- **Create a quiz:** Bebo has a quiz wizard for instant, original, and personal quizzes. These quizzes are often used to see how much your friends know about you.

✔ **Poll your friends:** Bebo also has a poll wizard to help you create multiple-choice questions. As people respond to the poll by selecting an answer, the results of the poll appear. Poll takers know immediately how much in or out of sync they are with your personal community.

✔ **Upload multiple photo albums:** Many online communities let you post a picture, but Bebo lets you post scores of images. Each album on a Bebo profile page is represented by a thumbnail photo linked to a larger album.

✔ **Create a blog:** Bebo has a section for ongoing online diaries called blogs (short for web logs) with links to the blogger's profile as well as a link to reply directly to the blogger.

To integrate Skype into its Web community, Bebo adds a field in the registration page for a Skype Name when you join. If you want to contact someone on Bebo, you can email, post comments, or skype directly from each profile page. Skype status icons are posted alongside on the Bebo user's profile page, so you know who is online and available to talk (see Figure 13-2). Of course, if your new friend decides not to authorize you to make this Skype connection, you'll just have to keep looking for another willing Bebo companion.

Figure 13-2:
The Skype status icon and Skype links on a Bebo user's Home page.

Close encounters of the Skype kind

The Skype Journal (www.skypejournal.com) is a community "newspaper" with a difference. The community this resource attracts (namely, skypers) has a direct pipeline to the journalists who write for the publication. Each writer has a byline with a Skype status icon next to his or her name. In fact, writers' names are direct links to a Web page with clickable icons for Skype calls, text messaging, Skype Voicemail, and even an old-fashioned phone number (suitable for a SkypeOut call).

The Skype Journal posts formal articles on all things Skype. Topics include gadgets, Web sites, software, controversies, anecdotes, and interviews. Each article is filled with hyperlinks (clickable connections to more information), a Listen to This Article link to a spoken version (as a QuickTime movie), and a TrackBack link that connects to previously published articles on a similar topic. In addition, every article has an Add Your Comments link for instant feedback on the article or comments on the topic. The Skype Journal archives the latest 200 posted reactions, which are accessible on any page of its Web site by clicking the "Skype Journal comment river." Wading through the "comment river" yields links to blogs, Web sites, and Skype Names and provides a handy link back to the referenced article.

Whereas the Skype Journal is a topical gathering place, the Skype Forum (forum.skype.com/) consists of a lively collection of discussion threads hosted by Skype itself. The Skype Forum publishes comments, questions, and stories related to personal, professional, imaginative, sentimental, economic, and technical issues. Everyone who signs into the Skype Forum can add a Skype Name (as well as an AOL, Yahoo!, or MSN username) to a profile, so it's possible to make direct contact with a kindred spirit around a hot Skype topic.

Joining Education Communities

Learning online, within a vibrant multilanguage, multi-age, and multicultural community that crosses time zones, political affiliations and geography, forces a change in educational habits. Joining a learning community means that the students have to "construct" their own learning by participating in live video debates, posting PowerPoint resources, contributing to threaded discussions, and honing written language skills by chatting with other students who may be speaking other languages. A new way of learning is emerging, shaped by the need for online learning collaboration and demanding written, spoken, visual, synchronous, and asynchronous participation by both teachers and students as a group and as individuals.

The addition of Skype to online learning has opened a powerful, live, face-to-face, and chat-to-chat arena with the following advantages:

✔ **Authentic conversations:** When you can learn a language by connecting to a native speaker, chances are you won't want to return to memorizing dialogues from books. The same applies to a conversation with an author or an expert in your field of study. The opportunity to meet, interview, and respond by including Skype in an online course connects the facts to real people in a very compelling, personal way.

✔ **Nontraditional feedback:** Traditional feedback, such as tests, grades, and marked papers often define how students learn in a classroom. Students study, take a test, and then wait for feedback in the form of a grade. New kinds of feedback experiences in forums, chat rooms, and live conferences provide an immediate response to students. Students not only react to ideas but also demand reactions from others, and they don't have to wait until the end of the term to figure out whether they've learned anything.

✔ **Unexpected collaboration:** School is sometimes a lonely, solitary pursuit. Students spend endless hours on the Internet doing research, reading texts, and completing homework. Online classes can exaggerate that isolation. Adding Skype conferencing, video connections, file transfer, and chats invites collaboration and academic give-and-take because Skype, by design, is a social place.

In the Moodle for learning

Many people take neighborhood continuing-education classes to meet kindred spirits. Now that our neighborhoods are online and have virtually no boundaries, no one has to settle for a class in word processing when Conversational Spanish would be preferable. In a brick–and-mortar school, the menu of classes is often limited. In a virtual school, the offerings can be almost endless. One of the most active online learning communities to fully integrate Skype is called Moodle, found at `www.moodle.org`. Moodle is a free, online course-creation software that provides quiz construction, grading, surveys, open forums, threaded discussion, and wikis (which are online references created, edited, and updated by the Web community). Anyone can download Moodle, host it on a server, load it with curricular materials, and run an online school. (For the curious, the word *moodle* derives from the acronym for Modular Object-Oriented Dynamic Learning Environment, which is useful mostly to programmers and education theorists. It's also a verb that describes the process of lazily meandering through something, doing things as it occurs to you to do them — an enjoyable tinkering that often leads to insight and creativity.)

World Class Schools

At one time, the PA (public address) system defined cutting-edge school communication. Now, of course, thousands of schools have high-speed Internet connections and the potential to communicate with the rest of the world through every available type of media. One organization, started by Michael Cunningham, a former principal who returned to a Texas classroom, reaches around the world with a Skype-based learning network run by teachers and students. Schools are connected through the World Class Schools portal (worldclassschools. 00politics.com) and are invited to join in formal debates, poetry slams, and ongoing foreign language exchanges. Some of the recorded video Skype sessions are posted on the Web site as QuickTime files along with scores of photos from the other areas of the world.

There is no cost beyond maintaining an Internet connection, which is often already installed in schools. Organizing school-to-school projects is easily arranged because the participants can talk to each other. Debate topics are posted and scheduled, and students prepare their topics according to rules sent out through Skype file transfer or email. Computer rooms are changed into living language labs as English-speaking students in Texas sharpen their Italian, Portuguese, or Swedish with their counterparts in Italy, Brazil, and Sweden. One extraordinary event paired students in the United States, Turkey, and Uzbekistan through Skype-enabled parliamentary debates.

World Class Schools takes advantage of available Skype tools to foster an active community. In a conference with a new school wanting to participate in one of these exciting programs, a veteran member can instantaneously send all his or her Skype contacts through file transfer. The result is an instant network of tech-savvy educators ready to launch student-to-student projects. The new list of contacts comes packed with information on location, interests, pictures, Web sites, and anything else an educator wants to share in a profile. All a teacher needs to do is click a Skype contact to begin to launch an international community event. Of course, all Skype contacts need to be authorized and your contacts must coordinate time zones, class compatibility, technical setups, and other organizational issues. But these are issues easily overcome, and the payoff in expanding the community of learners for your students is well worth the effort.

On Moodle, you can register and enroll in online courses. During registration, Moodle users can add their Skype Name to their profile, along with a list of various methods of communicating (see Figure 13-3). A dynamic combination of learner-initiated activities guarantees that nobody will fall asleep in the back of the virtual room. Take a quiz; give a quiz. Answer a poll; create your own survey. Make a movie and post it in Moodle class resources. When the mediator/teacher asks a question, Moodle users can now skype their answers through file transfer, or just speak up in class, if the instructor chooses to conduct class with Skype as the central communication tool.

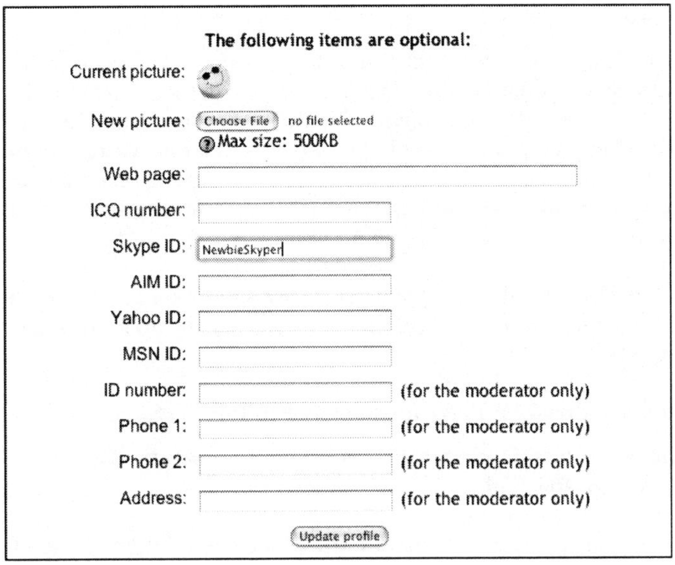

Figure 13-3:
Moodle
student
registration.

It's no surprise that Moodle has embraced Skype technology in its arsenal of learning tools. With Skype, a Moodle course enables students to

✔ Connect to similar classes across the globe and share ideas in a real-time mini-conference.

✔ Confer with the lecturer directly rather than through email or threaded discussions. You don't have to wait for a response when you can simply engage in a conversation.

✔ Exchange work on the fly through file transfer. Skype is ideal for educational collaboration.

✔ Form study groups "after class" by skyping your classmates.

Parlez-Vous Skype?

Language isolates communities from one another; studying languages connects them. The excitement of speaking another language goes a long way toward bringing extremely diverse cultures together. Probably the most natural use of Skype in education is to link students of foreign languages. After all, the best way to acquire a language is to speak it. The best way to speak a foreign language is with a native speaker, and Skype lets you speak with anyone around the world. As Skype technology takes hold, more and more language learning sites are emerging around the ease of using Voice over Internet Telephony, or VoIP. You can use Skype to acquire a new language on the following Web sites:

The Mixxer

Dickinson College created this site at www.language-exchanges.org as a free language exchange site. The Mixxer provides a search tool for finding language partners who use Skype. To join The Mixxer, register with an email address (which serves as your login name) and a password. Fill out a profile screen with your Skype Name, time zone, your native language, and the language you want to study. Your picture, age, and interests are optional information.

To let your Mixxer language partners know when you are available on Skype, you have to change your Skype privacy settings. To do so in Windows, follow these steps:

1. **Choose Start➪All Programs➪Skype to launch Skype.**

2. **Choose Tools➪Options➪Privacy and then select Allow My Status to Be Shown on the Web.**

To change your privacy settings on the Macintosh, follow these steps:

1. **Click the Skype icon in your dock to launch Skype.**

2. **Choose Skype➪Preferences➪Privacy and select Allow My Status to Be Shown on the Web.**

On Mixxer, you can explore the site and search for possible language partners Mixxer finds. For each prospective partner, Mixxer posts a name, a Skype Name, language, location, and local time. Mixxer also provides links to a user's Skype Profile and text chats, and posts Skype presence icons so that you know when someone you want to contact is online.

xLingo

xLingo (www.xlingo.com) is another free site promoting language exchange using a host of online tools including Skype. After joining xLingo using a standard sign-up screen to enter a username, email address, password, location, and language preference, you find numerous options for picking up a new language. Among them is to search for and connect to language partners through Skype. You don't need to search for any Skype contact information; instead, select a partner from the search list and click the name to pull up an xLingo profile. Each profile connects directly to Skype.

My Language Exchange

My Language Exchange (www.mylanguageexchange.com) is free, allows a partner search, and recommends using Skype to take advantage of its resources, although Skype is not integrated directly into the online software. My Language Exchange is a formal language-learning site with very specific methods. It promotes the Cormier method of language exchange, developed at the Club d'échange linguistique de Montréal (C.E.L.M. language school) in combination with Skype audio conferencing. With the Cormier method, you do the following:

- ✔ **Form small groups:** Gather no more than four people so that everyone gets a chance to speak. The ideal group contains equal numbers of speakers of each language (two English and two French, for example).

- ✔ **Find conversation partners who understand 40 percent to 70 percent of a conversation:** This is harder with a larger group but perfect for a Skype conference.

- ✔ **Speak one language half the time and then switch and speak the second language half the time.**

- ✔ **Use a timer to keep the group rotating languages:** Every Skype call window has a built-in timer, perfect for keeping the language flow.

---

eBay and commerce communities

Social networking is exploding in the world of e-commerce. As more and more people turn to the Web to buy, sell, swap, trade, and market goods, new mechanisms keep the Web economy responding to the power of a large community.

One of the biggest and busiest e-commerce sites, eBay, is an old hand at developing strategies to keep the community engaged and therefore keep businesses honest and competitive. eBay established rating systems to judge the quality of service offered by any firm or individual. The self-policing community establishes the reputation and identity of a business by voting on satisfaction. In this system, advertisements are less of a factor than the comments of fellow customers. The bid system is, in itself a group activity, establishing the worth of an item by the community of buyers rather than a cartel of sellers.

Community forums have grown up around sales items. Disneyana, comics, shoes, and a long list of item categories make a place for groups to discuss specific products and otherwise get to know fellow eBay members. An area for general discussion has names reflecting actual community social spots: the Front Porch, the Homestead, the Park, the eBay Town Square, and the Soapbox.

The acquisition of Skype by eBay extends the well-established culture of community. With this addition, three powerful programs work together. eBay is the commerce engine, Skype is the communications engine, and PayPal is the payment vehicle for this community. Skype enables buyers and sellers to communicate directly with each other. Imagine if you were looking for an antique and found one you liked on eBay. With Skype integrated into eBay, you can instantly have a chat with the seller, and because the item is expensive, you want to know everything about it. You may even want to pay through Skype as well. Talking over Skype is personal enough to get specific information from the person you are buying from and yet private enough for the buyer to feel a bit anonymous. A touch of anonymity takes the pressure off the buyer and gives him or her a little thinking space.

Skype and eBay are joining e-commerce and social networking to build a huge, interwoven and powerful community.

Skype is incorporated in this method by providing a simple, immediate, global way to speak online. In addition, you can use the Skype chat window to do the following: clarify new words and terms; use the Skype file transfer tool to share lesson plans (available on the My Language.com Web site); and use the Skype online status icon to let your study buddies know when you are online and ready to start class.

Skyping in All the Right Places

There may be thousands of people intrigued by the same hobbies you are, passionate about the same causes you embrace, or fascinated by the same collectibles you cherish. The Internet has a long history of bulletin boards, newsgroups, and so on, but Skype enables these communities, both large and small, to connect in a new way, invigorating and empowering each interest group with tools formerly too complicated or too expensive to implement.

Forming communities around special interests

Any interest group, club, or society can benefit from spreading Skype around, plugging in those microphones and talking to each other on a regular basis. Here are some types of groups already skyping away happily:

- ✔ **People with disabilities:** Individuals with mobility problems can join a group over Skype. For people with limited mobility, an online visit eliminates the difficulty of finding transportation, calling a paratransit bus, or getting into an adapted vehicle. Skype software can also be installed on computers with touch screens and special switches to enable alternative mouse access for individuals with limited physical movement.

- ✔ **People who communicate in sign language:** Individuals with hearing challenges can use the Skype chat window and file transfer as well as videoconferencing to communicate in sign language.

- ✔ **Musicians:** Musicians are beginning to experiment with Skype as a tool for remote jamming, collaborative composing, and performing.

- ✔ **Collectors:** All collectibles — stamps, shells, vintage toys —are a rallying point for a Skype group. Collectors love to talk, trade, and just bond with other collectors. Collectors can transfer Contact lists through Skype to share with fellow enthusiasts. You can even enjoy an active group if you live in a small town that is unlikely to have another passionate rock tumbler living in the neighborhood.

✔ **Gamers:** Gamers are used to online communities, which tend to be imaginary realms with richly costumed avatars. Some players are combining gaming with Skype, so you can converse with your opponent, discussing strategy and making a few friends outside the virtual arena.

✔ **Family Communities:** Families are reconnecting over the Web. Grandparents receive digital photo albums and email greetings, and they can view elaborate Web sites created for newborns. Most of this happens asynchronously, that is, not everyone is in the same virtual room at the same time. Skype can change that. With an easy conference call, Susan can conference in Mom and Dad in Florida, her sister in Flemington, and her brother in Newburgh. Everyone can talk, enter comments in the chat window, and turn on those video cameras for a real family gathering.

Skyping for a common cause

People always bond together to help others in need. Charities, victim services, outreach, and disaster relief all find representation on blogs and Web sites in addition to physical gatherings. Three key questions that organizations that ask are, How quickly can we help, How efficiently can we plan joint efforts, and Do we have the funds to get things done? Skype technology goes a long way toward helping in these areas in the following ways:

✔ **Cost savings:** Skype enables a significant reduction in long-distance telephone costs. If you skype through your computer, no cost is associated with that call.

✔ **Time savings:** Another great expense for relief organizations is time spent on personnel and travel. Eliminating even some air, rail, or rental car costs puts more in the coffers for those in need.

✔ **Planning and organizing that save lives:** One organization devoted to building shelters and housing for refuge and disaster victims has posted a Web-based Skype Me button for instant access to the central office. Of course, bringing all the members of a group into a common environment for Skype collaboration takes time, but after it is established, an enormous amount of coordination can take place in record time. With voice, chat, and file transfer available, many of the elements of a dynamic, timely, successful relief effort are in place. Planning is spent on how to help, not how to get to a meeting in order to talk about how to help.

Chapter 14

Skypecasting

*T*own Hall. These two words evoke a home-town scene, a designated place for members of a local community to gather, to speak up, and to raise issues. Holding meetings in a Town Hall is a particularly effective forum for people to come together and exchange ideas. Skype has its own Town Hall. Its designated place is cyberspace, and its home-grown community is made up of skypers. This Skype Town Hall is called *Skypecasting*.

This chapter shows you the ins and outs of Skypecasting, and you see how Skypecasting offers so much more than a traditional Town Hall.

Skypecasting: More than Just an Online Town Hall

Skypecasts are a way for people in groups of up to 100 at a time to congregate over Skype. Accommodating 100 people is a pretty nice leap from the Skype conferences of five or ten people that are covered in Chapter 10. The following sections give you just a taste of the full potential of Skypecasting.

As with a conventional Skype conference call, a Skypecast is totally free. Skypecasting works a little differently than a regular Skype conference call. The key differences are the following:

✔ **Participants join Skypecasts.** The conference host or organizer doesn't choose and call participants; instead, the participants choose a meeting on their own.

✔ **Skypecasts are public.** Skypecasts are listed on the Web and in a searchable directory within Skype. Anyone with a Skype Name can join a Skypecast.

✔ **Skypecasts can accommodate a crowd.** Skypecasts can be as small as one or two people and as large as 100.

✔ **You can join Skypecasts only on the Internet.** You cannot join a meeting by calling in from a regular telephone.

✔ **Skypecast hosts have special controls.** Participants in a Skypecast can request the microphone so that all the participants can hear what she or he has got to say. The host can individually mute or unmute the microphone of any participant.

The current cutoff for audience size in Skypecasting is 100 participants. If you want to give people the choice of connecting into a large conference either through Skype or calling from a regular phone, look at High Speed Conferencing (see Chapter 12). With this option, you can convene up to 500 people.

Skypecasting of the people, for the people, by the people

Skypecasts are for anybody. Anybody can join a Skypecast. Anybody can start a Skypecast. In contrast to members of other conferences, Skypecast members don't have to be dues-paying participants. They don't have to join a club. They are free to participate with no restraint except good manners.

The model for a Skypecast doesn't have to be a town hall. Often it is a soapbox. It can be whatever you want it to be. You can

✔ Start a club.

✔ Give a lecture or even a series of lectures.

✔ Create a story circle.

✔ Lead a guided meditation.

✔ Give a music lesson.

✔ Organize an impromptu class. Combine your Skypecast with Webinar (a Web seminar) software to deliver a PowerPoint or any other presentation.

✔ Get on a virtual soapbox and start talking. Anyone can join you. If you are compelling, interesting, or just plain unusual, people will stick around and listen.

The authors of this book set out to Skypecast a discussion on the basic setup and inner workings of Skype. Such a discussion brought together a small community of people with an interest in meeting other skypers, meeting the authors of a book, getting some of their questions answered, and having a soapbox for suggestions. The authors, on the other hand, wanted to meet the community of readers of books on technology, or the community of early adopters of new technologies.

You can initiate a Skypecast from any location. You can be lounging at home, working at the office, or enjoying your beachfront property (assuming that you have a broadband Internet connection). Skype connections have been made in remote places such as Mount Everest, so if you wanted to climb that mountain and Skypecast the tale of your journey, go right ahead. Most of us, however, have things to share from locations less exotic than Mount Everest.

Finding a Skypecast

With so many topics available, you may think that finding the Skypecast you want to join would be difficult. Fortunately, you can find Skypecasts in a few different ways:

✔ **Go to the Skypecast Preview Directory.** Go to the Skypecast home page (`https://skypecasts.skype.com/skypecasts/home`). On this page, you can find

  • Featured Skypecasts

  • Skypecasts scheduled over the next four weeks

  • Skypecasts about to start

  • Skypecasts already in progress

  You don't need to run Skype when you view these listings; they are regular Web pages. You do need to have Skype installed on your computer to join a Skypecast, however.

✔ **From your SkypeLive page on your Skype client, view a list of Skypecasts already in progress or that are about to start.** If you are using Windows and have Skype 2.6 or later, you have a feature called SkypeLive (see Figure 14-1).

✔ **Find out about Skypecasts posted on a blog or Web site.** Blogs and Web sites that post links to Skypecasts on specific topics can be kept up-to-date as Skypecasts are added and revised.

✔ **Find out about Skypecasts listed on RSS feeds.** Web pages commonly have RSS links.

Figure 14-1:
Skypecast
listings on
the Live
page.

Aside from browsing these lists, you can search for a Skypecast by topic or keyword (sometimes referred to as a tag), by the Skypecast title, or by the Skype Name of the Skypecast organizer. You enter the search term and then click the Search button in Find a Skypecast on the Skypecast Home page (`https://skypecasts.skype.com/skypecasts/home`).

Joining a Skypecast

If this is your first time to join a Skypecast or a similar type of online forum that allows you to speak up and be heard, then you're sure to have fun. The steps involved in joining a Skypecast are easy:

1. **Make sure that you have the most recent version of Skype.**

   Make sure that you have Skype software (version 2.5 or later for Windows and 1.4 or later for the Mac). New Skypecast features continue to be introduced. For instance, the Skype Live feature, which displays current Skypecasts directly within Skype, can be found on version 2.6 or later for Windows, but version 2.5 does not include the feature. To take better advantage of new Skypecast features, get the latest version of Skype for your platform.

2. **Find a Skypecast at https://skypecasts.skype.com/skypecasts/home.**

   Note the date and time that the Skypecast is scheduled to start.

3. **When it's time to join the Skypecast, browse the list of Skypecasts posted on the Skypecast home page (see Figure 14-2) and click the <u>Sign in to Join</u> link for the Skypecast of your choice.**

Figure 14-2:
Choosing a
Skypecast.

4. **When prompted, enter your Skype Name and password.**

   With Skype 2.5 for Windows and 1.4 for Mac, a separate program is launched to connect to the Skypecast. Later versions of Skype may integrate Skypecasting directly in the Skype program instead of launching a separate program.

Organizing a Skypecast in 5 Minutes or Less

Starting your own Skypecast is as easy as going to the Skypecast Web site, finding the link to create a Skypecast, and entering the information for the Skypecast you want to hold.

If you are adventurous, follow these steps to set up a Skypecast right now. You don't need a lot of lead time to Skypecast, just log on and speak up!

1. **In your Web browser, enter the address**
   https://skypecasts.skype.com/skypecasts/home.

   Note the *s* in "https"; this means it's a secure site.

2. **Click the <u>Create a Skypecast link</u> (see Figure 14-3).**

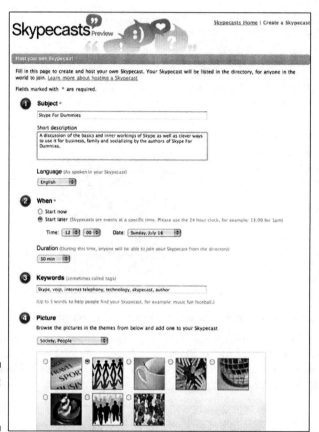

Figure 14-3:
Creating a
Skypecast.

3. **Enter a meeting subject.**

   Your meeting subject should be a brief one-line description. You must enter a subject to set up a Skypecast. Choose a title that will entice people to join and be easily picked up on searches.

4. **Enter a short one or two paragraph description.**

   The description is also a required item.

5. **Select the primary language that is to be spoken.**

6. **Enter the time and date for your Skypecast.**

   This item is required. You can have your Skypecast start immediately without any advance scheduling, but most skypers schedule their Skypecasts in advance. You can schedule the Skypecast for any day up to four weeks out into the future. You need to specify the time of day in terms of a 24-hour clock cycle (as opposed to a 12-hour cycle).

Remember, Skypecasts are global. Pick a time that's not only convenient to your local community but also accommodates a time zone that contains your likely audience. If you have a Skypecast at noon in Boston, you can expect some Californians to join in at 9 a.m., some Londoners to join in at 5 p.m., and a resident of Moscow to join at 9 p.m.

7. Select the duration.

During this time, anyone will be able to join your Skypecast from the directory. You have a choice of setting the duration from 15 minutes up to 5 hours.

If you don't want late stragglers coming into your Skypecast, select a shorter period of time such as 15 minutes or a half hour. Keep in mind, though, that if a person who joins a Skypecast has a problem with his or her computer, that person may be forced to leave the Skypecast and can't rejoin the Skypecast after the allotted duration expires.

8. Enter keywords for your Skypecast.

You can enter up to five keywords, or *tags*. Entering keywords is optional, but we recommend adding them because they help people find your Skypecast more easily and better understand the topic of your Skypecast.

Try to vary the keywords rather than restrict them to the same topic.

- Browse the picture themes to find a picture that quickly communicates your topic. The picture themes are as follows:

  - Arts, Movies, Music
  - Business, Meetings, Finance
  - Flirt, Love, Romance
  - Gaming
  - Health, Fitness
  - Holidays, Vacations, Travel
  - Internet Chat
  - Mothers, Parenting, Family
  - Science, Technology, Ideas
  - Society, People
  - Sports

9. Preview the listing, review and accept the Skypecast Community Guidelines, and click the Agree button.

After you agree, the Skypecast is scheduled.

Building a fan base

Building a community, or a fan base, begins with finding a good topic or set of topics for your Skypecast that inspires you and your audience to meet online. Here's a cross-section of ideas to help you get started:

✔ Create a daily Skypecast to get on your soapbox in an election.

✔ Skypecast your latest musical compositions to create a fan base.

✔ Gather to watch the Super Bowl over an open Skypecast with a community of fans rooting for both sides.

✔ Host Skypecast sessions of a public conference. Invite attendees to create Skypecasts to develop the ideas raised at the conference.

✔ Interview a popular author and Skypecast the event. Follow up with a regular Skypecast book club.

✔ Host a regular movie discussion group.

✔ Create a forum for parents of children with disabilities to discuss, share stories, and support in a nonthreatening, information-rich arena.

✔ Start a technology users discussion group for a common software program such as Excel, Skype, Photoshop, or Final Cut Pro.

Between the time you schedule a Skypecast and the time it runs, you can edit the conference description. For example, you can reschedule and set it to an earlier or later date, or you can revise the Skypecast description. If necessary, you can cancel the Skypecast.

As an added bonus of hosting Skypecasts, you can take the description of a Skypecast you recently organized and reuse its description for additional Skypecasts.

Choosing a topic for your Skypecast

Choosing a subject for your Skypecast is easy. Pick something you are interested in, from collecting baseball cards to baking the ultimate chocolate chip cookies; from brushing up on your Japanese to watching the Super Bowl. If you are an expert in all these areas, feel free to make each one a Skypecast. You can host as many Skypecasts as you want, and it's totally free.

When you create and host a Skypecast:

- ✔ **Choose a title that sells itself.** Creating your Skypecast listing is an opportunity to let your prospective audience know what they are in for. A good title explains why someone should join your Skypecast and what you'll be talking about. Some examples of existing Skypecast titles that give you a hint as to the topic and type of discussion:

  - I'm a Brazilian, Learning English

  - Poker School Online Training Tournament

  - Creative Real Estate Investing

  - Survival Strategies of the Selfish DNA Programs

- ✔ **Think carefully about the keywords.** Keywords, or tags, are words or phrases that you add to help users to find your Skypecast by searching the directory. You have an opportunity to add keywords in a special field when you create a Skypecast. You can also add keywords after the Skypecast is posted. Keywords for a Skypecast titled "Financial Markets" are "dollar, euro, actions, stock, finance." These keywords also determine on which syndicated listings your Skypecast appears. Keywords help users find your Skypecast and ensure that your Skypecast gets the best possible promotion.

- ✔ **Promote your Skypecast.** Send the Skypecast link to your friends, family, and co-workers in an email. You can also add a simple HTML snippet to your blog or email signature block to remind your fan club (or future fan club).

- ✔ **Don't leave your audience in the lurch.** Note the time and day you are to host a Skypecast and be ready and waiting for your audience. A skyper entering an empty Skypecast will turn around and leave.

  Decide what kind of host you want to be and what kind of Skypecast you want to run. You may want an open discussion or a one-way broadcast with questions at the end.

  Use Skypecast microphone tools strategically. After your Skypecast begins, as a host, you can mute or unmute the microphones of some or all of your participants (be careful; this may become habit forming). Muting participants' microphones actually increases call quality for everyone in the Skypecast, so keep that in mind when voices begin to crackle. Participants can request the microphone by clicking a button available to them. The host assigns a speaker, passes the microphone around, and keeps the quality and flow of the Skypecast high.

  Don't hesitate to be a bouncer! If you have an unruly member bent on mayhem, toss that person out! Skypecast controls allow you to remove uncooperative participants.

Getting the Word Out

Your conference is automatically posted in a directory of current and upcoming Skypecasts at `https://skypecasts.skype.com/skypecasts/all` (see Figure 14-4).

Figure 14-4:
Featured
Skypecast
conference
listing.

There are several ways for people to learn about your Skypecast:

✔ Scheduled Skypecasts are listed in the Skypecast directory on the Skype Web site. Current and ongoing Skypecasts are also displayed on the Live page of Skype applications starting with version 2.6 for Windows. Both the Web listing and listing in the client are automatic and don't require action on your part.

✔ Listings can be posted to a blog using TypePad.

✔ Listings can be posted on any Web site.

✔ Listings can be posted on RSS Feeds.

Automatic directory listings

Skypecasts are listed in the order they take place. Because accommodating the various time zones complicates listing the time that each Skypecast will start, the directory posts a notice such as *Starts in 4 hours and 55 minutes.* This way, whatever time zone you are in, you know when to log on and join in.

Placing a Skypecast Widget on a blog

Another way to get the word out about your Skypecast is to add a Skypecast Widget on a blog. One company, TypePad (www.typepad.com), has partnered with Skype to connect the world of blogging with the world of Skypecasts. As a TypePad blogger, you can create a Skypecast and then add a special Widget to your blog. People reading your daily blog can click the Widget to join your Skypecast directly from inside TypePad. Your community of blog readers is a likely audience for your Skypecast adventure, so embedding a link in your blog is the best way to let them know when you will make your broadcasting debut.

To add a Widget to your TypePad blog:

1. **Log onto https://skypecasts.skype.com/skypecasts/skypecast/ widget.html.**

2. **Select a radio button for a featured Skypecast, your own Skypecast, or a tagged Skypecast.**

3. **Click <u>Install Widget on TypePad</u>.**

   If you select a tagged Skypecast or your own Skypecast, you will be directed to type your Skype Name in the field provided. Skype searches for the Skypecast you want to promote.

4. **Click the Install Widget on TypePad button.**

   The TypePad login page appears.

5. **If you have an account, enter your name and password.**

6. **Select the blog on which you want the Widget to appear.**

7. **Preview your Widget.**

8. **Confirm your choice.**

Placing a link to your Skypecast on your Web site

After you schedule a Skypecast, you can find its meeting ID number and place a link to it on your Web site. The link takes the form of the following:

```
https://skypecasts.skype.com/skypecasts/skypecast/detailed.html?id_talk=15783
```

The only portion of the link that differs from Skypecast to Skypecast is the `id_talk` number appearing at the end of the link. You can place this on your Web site by inserting the following snippet of HTML code into your Web page.

```
<a href="https://skypecasts.skype.com/skypecasts/
         skypecast/detailed.html?id_talk=15783"
         target="_blank">My Skypecast Title</a>
```

The `target="_blank"` portion tells the browser to open a new Web page when clicking the link.

To find out the `id_talk` number of your Skypecast, go to the general Skypecast directory listing and find your Skypecast. Click the link for your Skypecast. When the browser opens to your Skypecast details, look at the URL displayed in your browser. The digits appearing at the end of the URL is your `id_talk` number.

There are ways to insert into your Web site a listing of all the Skypecasts for a particular kind of search, such as all Skypecasts hosted by a specific Skype ID. This is technical and involves JavaScript. By the time you read this, someone, possibly Skype, may already have an HTML code snippet maker that automatically constructs all this geeky HTML and JavaScript code. Look to our site www.skype4dummies.com for up-to-date information on this. In addition, you can add an RSS feed to your Web site without any complicated coding.

Adding a Skypecast RSS feed

Placing a hardwired link to the Skypecast details on your Web site is easy enough, but if you hold multiple Skypecasts, it would sure be nice to list every Skypecast in which you are the host.

Nowadays, people are starting to use RSS feeds to tap into information that gets regularly updated. You can place an RSS link on your Web site. The RSS link appears as follows:

```
http://feedsskypecasts.skype.com/skypecasts/webservice/fee
         d.html?user=seasonedskyper
```

What is an RSS feed?

It can be tiring to constantly check for changes and additions to a Web site. To help keep up-to-date and follow a particular topic, you can use an *RSS feed*. RSS has many aliases. One of them that is easy to remember is *Really Simple Syndication.*

RSS is a sort of news feed that makes it easy to get regular updates on a given topic. RSS is not a product or software. Rather, it is a way to publish information in a format that many different kinds of computer programs can read. People who are making information available for distribution on the Internet don't have to worry about how the information is presented on every piece of hardware and operating system on the planet.

Some Web browsers can automatically read the content of an RSS feed. Other Web browsers let you choose an RSS Reader extension to add to your browser.

In your link, change the name `seasonedskyper` to your Skype ID.

If you want to show the listing of all Skypecasts having *business* as one of the keywords or tags, you can create a link like the following:

```
http://feedsskypecasts.skype.com/skypecasts/webservice/fee
          d.html?tag=business
```

If you want to show all currently featured Skypecasts, you can include a link to the following:

```
http://feedsskypecasts.skype.com/skypecasts/webservice/fee
          d.html?featured=1
```

Moderating Your Skypecast

The day is here. Your Skypecast is due to start in five minutes. What do you do?

1. **Sign on to join the Skypecast as you would if you were a participant.**

   The Skypecast conference window appears. Your name is on the top, listed as the host. As people join in, Skype Names appear in a list. A little timer is ticking away, letting everyone know how long the conference has been on the air.

2. **Introduce yourself and your topic. Start talking!**

3. **Invite participants to speak.**

   As the host, you are in control of the microphone. You can mute everyone and give out the microphone only as members request it, or you can unmute everyone and have a free-for-all (see Figure 14-5). If you are host-ing a sing-a-long or cheering a soccer team, you might want to open the mikes up full throttle.

If anyone gets too rowdy, you always have the power to eject a person from the Skypecast. After all, every community, even a virtual one, needs a few rules.

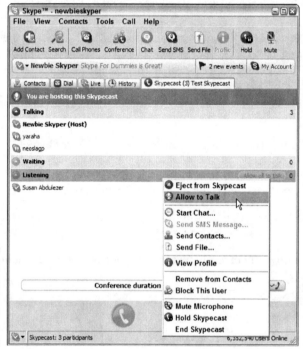

Figure 14-5:
The Skypecast conference window.

Your new community is launched. You can follow up with other Skypecasts, start a related forum on an existing Web site, or just maintain contact with some new Skype buddies. Skypecasting will change as new features are added and conferences become more tightly integrated into your regular Skype client. Your Skypecasts may become

✔ Private events

✔ Invitational events

✔ Charity fundraisers

✔ Political caucuses

✔ High school reunions

If you are an astronomer with an active stargazing site, you may soon be able to post a direct link to a Skypecast from your Web site. The addition of video into the Skypecast mix is a logical next step and, if developed, would multiply opportunities for a rich forum available at little or no cost to everyone. It is a remarkable tool for connecting people, and, although the look and feel of the interface may not be the same a month from now, the basic utility of Skypecasts as a community builder will remain the same.

Making Your Skypecast Everlasting

Skypecasts are live events. When they're over, they're over. However, sometimes you want to preserve the event. Some of the ways, and reasons, to save a Skypecast are as follows:

✔ **Transcribe your Skypecast:**

- Extract quotes for an article or research paper.

- Generate captions for a video presentation of your Skypecast topic.

- Create a transcript for reference. This is a good technique for, say, marketing research, developing a troubleshooting guide for software, and harvesting unique phrases, idioms and terms for a short story.

✔ **Save the Skypecast discussion by recording the audio:**

- Extract audio quotes for a PowerPoint presentation.

- Import the audio track into a video-editing program to use as an informational soundtrack.

- Create a podcast.

Tools for preserving your Skypecast

You can save Skypecasts as audio files or text documents by using a variety of software and hardware tools. We've included some specific tool suggestions. This is nowhere near an exhaustive list; but it's a start.

Depending on the jurisdiction of where you are recording the Skypecast, there may be legal requirements you need to adhere to that require you to inform your participants that the session is being recorded. Please be certain to inform the participants of your Skypecast that you are recording the event. It is both courteous to your guests and it may be legally required!

Saving your audio by using hardware tools

To save the audio of a Skypecast with hardware tools, do the following:

1. **Plug a USB speakerphone into your computer.**

   A speakerphone such as the Polycom unit described in Chapter 11 serves the need well.

2. **Set up a traditional tape recorder, a mini-disc recorder, or a flash-based recording unit to capture the audio.**

   If your recording is digital, you should be able to transfer your audio recording directly to your computer. The way you move the file may differ depending on your recording device and the manufacturer's instructions. If your recording device is not digital, we recommend that you consider acquiring a digital recorder; they are rapidly becoming more affordable.

    Some digital recorders come equipped with a "T-bar" stereo microphone. This type of microphone enables you to position the microphone between yourself and the speakerphone so that a stereo recording of the Skypecast has good sound separation.

Saving your audio by using an analog device

If you have a recording on a nondigital (analog) device such as a cassette recorder, you can:

1. **Attach a cable from the audio-out line of your recorder to the audio-in line of your computer.**

2. **Select the Record button of your Windows XP Sound Recorder (choose Start⇨Programs⇨Accessories⇨Entertainment on your computer).**

3. **Click the Stop button upon reaching the end of the audio playback.**

Note that files are saved as WAV files, which are much larger than MP3 files. An hour's recording may consume close to a gigabyte of disk space.

Saving your audio by using software tools

To save the audio of a Skypecast with software tools, use the following:

✔ Pamela or Skylook recording software for PC (see Chapter 9)

✔ Call Recorder for the Mac (see Chapter 11)

Transcribing Your Skypecast Recordings

So, you're busy recording anyone who will let you capture a conversation, and a folder is filling up with MP3s of interviews, meetings, stories, and plans on every subject. Some of those audio files are a great source of information to have in written form. For example, much of the research done for this book exists as recorded Skype calls preserved as audio files. But, as much as we would like to play our audio recordings for you, we have a few practical requirements. We are limited to paper and ink. That means that we have to convert what we hear to what you read.

Listening to hours of Skype recordings while typing a transcript is a huge job made easier by a wonderful software program, Transcriber (see Figure 14-6), available at no cost from `http://trans.sourceforge.net/en/ presentation.php`. This software is available for download for Windows, Mac, and Linux.

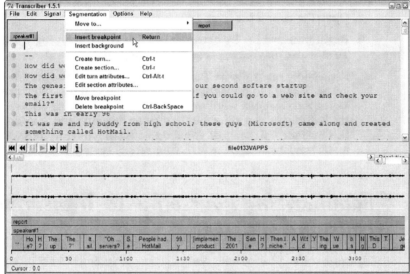

Figure 14-6:
The
Transcriber
program.

Transcriber's features allow you to

- ✔ Import MP3 files.
- ✔ Create labels with the names of individual speakers.
- ✔ Create new sections when there is a change of topic.
- ✔ Segment audio portions and sync them with written files.
- ✔ Zoom and telescope the timeline.
- ✔ Incorporate the names of speakers in the timeline.
- ✔ Export transcription files as text or XML (getting not only the text but also the time codes).

Transcriber shaves hours off the task of transcribing long audio sessions, organizing text by topic and speaker, and publishing the transcript in print or on the Web.

Getting started with Transcriber

Transcriber has many fancy features. But just a straight transcription, in many cases, gets the job done. Here is the two-minute tutorial to get you quickly up to speed:

1. **Launch the software.**

   You are prompted to locate your digital recording (typically, an MP3 file).

2. **Open the sound file.**

3. **Begin transcribing by pressing the Tab key and typing what you hear. To stop the playback, press the Tab key again.**

   Repeatedly pressing the Tab key starts and stops the playback. As you get acclimated to transcribing, you will find that you often press the Tab key at the start of a word or phrase, type the word or phrase, and press the Tab key as you are about to resume listening and transcribing. In this manner, you can just about do nonstop typing while slowing the audio track to keep pace with your rate of typing.

Familiarizing yourself with Transcriber's bells and whistles

Transcriber has a ton of bells and whistles, including the ability to mark your transcript with annotations on background noise, include music, identify different speakers, and even overlap speakers for any segment of your recording. Here are some tips for using the more advanced features of Transcriber:

- ✔ To rewind the audio playback a few moments, press Alt+←. This feature is useful when you are trying to transcribe something difficult to hear or when the person is talking at 90 miles an hour.

- ✔ To advance the playback to a later point in the timeline, press Alt+→. There are bound to be moments where there is a pause or natural break. Press the Enter key and continue typing.

- ✔ Every time you press the Enter key, you insert a new segment in the timeline, and the text window where you are transcribing inserts a new line, starting with a bullet point. Your timeline at the bottom of the window has dividing lines between segments. To move back and forth between segments, press the up arrow or down arrow key. In this manner, you can swiftly move to just about any portion of a long recording and rework the text for a particular segment.

✔ If you want to split a segment into two smaller ones, position your cursor in the text at the place you want to make the split and then press the Tab key to begin the audio playback for the segment. When you get to the moment in the audio playback when the segment should be made, press the Enter key to insert the segment.

✔ To join two segments into one, press Ctrl+backspace. Every now and then, save your file. The file that gets saved has not only a text transcription but also a time coding for each of the segments. This information makes it easy to create captions for audio and video recordings.

✔ If you have a long recording such as an hour in length, want a word-for-word transcription, and need to have all this done quickly, make three copies of your audio file and distribute them to three people. Have one person transcribe the first 20 minutes of the recording, another person transcribe the second 20 minutes, and a third person do the remaining portion. The three transcriptions can easily be merged.

✔ As a practical matter, if you need to do a word-for-word transcription and are reasonably comfortable with the software and the typing, plan on spending two-and-a-half to four times the audio recording time for creating the actual transcription.

This software has tons of features. Just following the steps here will get you off to a good start. Happy transcribing!

Podcasting Your Skypecast

You can preserve your Skypecasts and promote your ideas by hosting a Skypecast, recording it, and turning it into a podcast. Podcasts make your Skypecast ideas portable. Post your podcast on the Web, and anyone with a portable player can download and play your newly transformed Skypecasts.

The word *podcast* is a combination of *iPod* and *broadcast*. Podcasts are audio and video files published on the Web and downloadable onto iPod players through the Apple iTunes software. Podcasts are becoming a popular way to create personal "radio" programs. You can podcast on a regular basis, create a fan base, and even have listeners subscribe to your show to automatically download to iTunes as soon as you go on the air.

You can turn the MP3s of your Skypecast into a podcast in two ways. One way is to make an RSS feed that points to the MP3 file. This is the quick-and-dirty way. Everything in your unedited MP3 recording gets converted into your podcast, including long stretches of silence, background noise, and people fumbling around with their microphones and saying "Hello, can you hear me now?" You can't create lead-in music, include photos that get displayed on your browser, or make your iPod capable of displaying pictures.

Fortunately, taking the leap from the rudimentary transfer of it all to a mini-production system complete with studio-style music lead-ins, sound effects, visual effects, and captioning is easily done and inexpensive. Better yet, the tools for the mini-production studio are intuitive and anything but geeky. The tools we have in mind and recommend are found in the Apple iLife suite of software.

Creating a basic podcast on Windows or Mac

Skypecasts are great sources of content for new podcasts. In broad strokes, to turn a Skypecast into a podcast, you must do the following:

- ✔ **Record the Skypecast:** Use Call Recorder for the Mac, Pamela Pro for the PC, or Audio HiJack for either platform to capture the audio.

- ✔ **Edit your audio:** Editing programs such as Audacity for Mac or PC, or GarageBand on the Mac, enable you to cut, paste, and arrange your audio files for your "show."

- ✔ **Create an RSS feed:** RSS stands for Really Simple Syndication. When you create an RSS feed for your audio file, you are making your podcast available for subscribing and downloading from a Web page. A variety of sources, such as Blogger (www.blogger.com) and FeedBurner (www.feedburner.com), can create the RSS feeds for you.

- ✔ **Add your podcast to your blog or Web site:** To publicize your Skypecast-turned-podcast, you can submit the RSS feed page to Apple through iTunes. If the content is acceptable for posting, iTunes puts it into its podcast directory. Your Skypecast is available for downloading.

Using FeedBurner to create a podcast

To use a tool such as FeedBurner, you need to have a blog or Web site set up and available, so you have to flex your geek muscles a bit before using this tool. If you are computing on a Macintosh, you can skip directly to the podcasting with iLife steps later in this chapter. To create an RSS feed with FeedBurner, follow these steps:

1. **Set up an account (you can start with a free account).**

   • Enter a username, password, and email address.

   • Optionally, you can enter a secret question and answer, which is useful in case you forget your password or require some fundamental change to your account.

   • Click the Sign In button.

2. In the screen labeled My Feeds, enter the URL of your blog or feed page.

This URL needs to be the actual page you are using for your podcast (see Figure 14-7).

Also, select the I Am a Podcaster option. This option instructs FeedBurner to use SmartCast, which makes your feed podcast ready.

3. Click Next.

A Welcome page opens (see Figure 14-8), in which you can enter your feed title and feed address.

4. Give your feed its title and feedburner.com address.

FeedBurner guesses your feed title from your blog and allows you to set the feed address (see Figure 14-8), which always starts with `http://feeds.feedburner.com/` and is followed by an editable word based on your podcast title or podcast theme. An example is *Skypecasting to Podcasting* for the title and `SkypecastingToPodcasting` for the feed address portion. The resulting URL becomes `http://feeds.feedburner.com/SkypecastingToPodcasting`.

5. Click the Activate Feed button.

FeedBurner posts your feed, which is now live for anybody to view and subscribe to as the RSS feed.

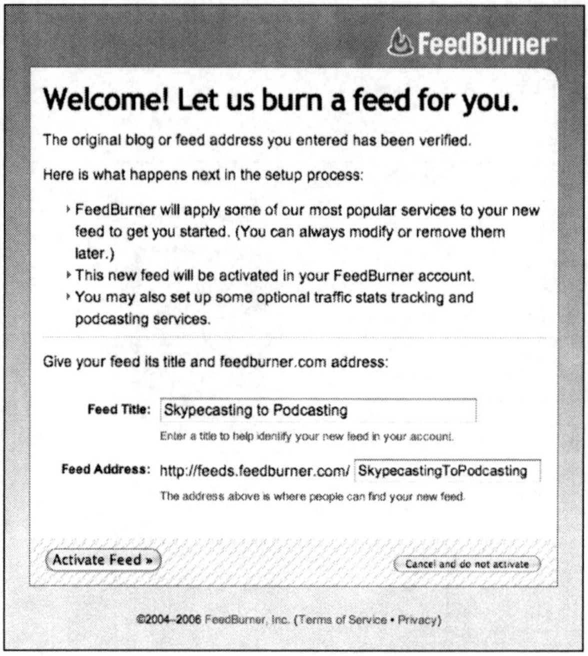

Figure 14-8:
Setting your
podcast title
and feed
address on
FeedBurner.

Using iLife to create a podcast using a Mac

Using Apple's iLife '06 suite of software and the .Mac portal, you can publish the audio file of your Skypecast quickly. These tools work together:

- ✔ **GarageBand.** This is the iLife sound editing software. GarageBand has a special track to add chapter markers, bookmarks, and graphics files to an audio file for podcasting.

- ✔ **iTunes.** This is the iLife music library. Store your recorded Skypecast files here by simply clicking the MP3 file to open it. The default player is iTunes. After being opened in iTunes, it is stored in the iTunes library, which makes it handy for you to retrieve the Skypecast MP3 file later.

- ✔ **iPhoto.** This is the iLife photo and image library. If you want to put any artwork or photos in your podcast, park them in iPhoto by simply dragging the images onto the iPhoto icon. If you are importing photos for your podcast, your digital camera will automatically open iPhoto when attached. Click Import to bring the photos into your library.

- ✔ **iWeb.** The iLife Web page editor automates the creation of an RSS feed. This is major. RSS feeds are complicated even if you are a geek!

- ✔ **.Mac Account.** The Apple email and Web page services are completely integrated with iLife. Publishing a Web page with a podcast is simply a matter of pressing a few buttons.

Podcasting your recorded Skypecast is fairly straightforward, unless you want to get fancy and add artwork, jingles, and sound effects. Be aware that all your media-sound files, images, and photos must be stored in iLife software so that you can access them easily. The basic steps are as follows:

1. **Click the GarageBand guitar icon in your applications dock to launch GarageBand (see Figure 14-9).**

   Click New Podcast Episode from the list of new project types.

Figure 14-9:
Creating
a new
podcast
episode
with
GarageBand.

2. **Title your GarageBand podcast and click Save.**

   It's a good idea to make a separate folder for your podcasts, especially if you plan to do a series and save all the episodes in one place.

   GarageBand opens (see Figure 14-9). Review the three main areas on the screen:

   - Tracks are located in the upper-left pane.
   - Podcast artwork and chapters are in the lower-left pane.
   - The Media Browser is on the right. Along the top of the Media Browser, you see three tabs for Audio, Photos, and Movies.

3. **Click the Audio tab in the media browser and select the MP3 file you want to podcast from the iTunes library displayed within the GarageBand media browser.**

4. **Drag the MP3 file into a GarageBand audio track.**

   You can change the name of any track by selecting the text and entering a new name. You don't have to add audio to every track. GarageBand provides five tracks, labeled as follows:

- **Podcast Track.** This track is exclusively for artwork that is displayed while the podcast is played. The other four tracks are for audio.

- **Male Voice:** Suggested track on which to record a Male voice.

- **Female Voice:** Suggested track on which to record a Female voice.

- **Jingles:** Suggested track on which to record a jingle or tune. You can create jingles and sound loops that are copyright free.

- **Radio Sounds:** Suggested track on which to record radio sounds.

5. **Click the Photos tab in the GarageBand media browser and, from the iPhoto library displayed, drag a photo onto your Podcast track.**

   Drag the photo onto the blank Episode Artwork window.

6. **Click the information button (it shows a circle with an *i* inside) to add Episode information (see Figure 14-10).**

   The available information includes the following:

   - **Title.** Add the name of your episode.

   - **Artist.** Enter the name of the podcaster.

   - **Description.** Write a clear description of your podcast. This is an opportunity to pique a listener's curiosity.

   - **Parental Advisory.** GarageBand also provides a Parental Advisory drop-down list. Select None, Clean, or Explicit.

7. **Choose Share⇨Send Podcast to iWeb.**

   GarageBand converts the file to a podcast. When it's complete, the sign-on screen to your .Mac account automatically opens. Sign on to your .Mac account (see Figure 14-11).

8. **Select a template for your Web page in iWeb.**

   Choose from among the different themes that appear in a scrolling list with items such as Road Trip or Travel.

9. **Apply your template to Podcast.**

   You have two template choices, Blog and Podcast. Select Podcast and click the Choose button. iWeb creates a Web page with your podcast. You can make some changes on this screen before you publish your podcast.

10. **Click the title Podcast and change the text to your own title.**

    You can also select any text in the description and change it in iWeb.

11. **Click Podcast in the iWeb site organizer.**

    At this point, you can preview the page as it will be seen on the Web. Change the images and text to reflect your podcast.

Figure 14-10:
Adding your podcast picture and description in GarageBand.

Figure 14-11:
If you don't have a .Mac account, you can sign up for one on the spot.

12. Click Publish.

This puts your Podcast on the Web by uploading the Web page to your .Mac Web site. iWeb includes a Subscribe button on your Podcast site; anyone can click Subscribe and automatically receive newly posted podcasts from your site (see Figure 14-12).

Skypecasts transformed into podcasts can easily be played on iPods, which is the MP3 player of choice for more than 40 million people. Any artwork you include when you create the podcast in GarageBand is displayed on the iPod screen.

Figure 14-12: Your published podcast.

Submitting your podcast to iTunes

Now that you have your podcast on a blog or Web site, the time has come to let the world discover you. The iTunes music store is *the* podcast directory.

To submit your podcast to iTunes, follow these steps:

1. Launch iTunes.

iTunes is a free download for Windows or Mac and is available at www. apple.com/itunes/download.

2. **Click the <u>Music Store</u> link.**

   On the Source panel on the left is a variety of items to pick from, including Library, Podcasts, Videos, Party Shuffle, Radio, and Music Store. To submit your podcast, click the <u>Music Store</u> link.

3. **Click Podcasts in the Inside the Music Store pane.**

4. **Click either the large Submit a Podcast button or the <u>Submit a Podcast</u> link.**

5. **Enter the URL of your podcast's RSS feed.**

   This URL is likely to have an `.xml` suffix. Click the Continue button, and the screen shown in Figure 14-13 appears.

Figure 14-13: Submitting your podcast to the iTunes podcast directory.

6. **Sign in to publish your podcast.**

   You need an Apple ID or AOL screen name to publish your podcast. If you don't have an account, click the Create Account button (see Figure 14-14).

   Creating an account is free, but you must provide credit card information. This allows you to act on impulse and instantly buy something at the Music Store (and get charged on the spot!).

7. **After you create your account, or if you already have an account, click Continue.**

8. **Set the Category, subcategory, and language for your podcast and click the Submit button (see Figure 14-15).**

Figure 14-14:
Signing in to publish your podcast on iTunes.

Figure 14-15:
Creating iTunes categories for your podcast.

An automated email response is sent to you and informs you that the podcast feed has been received and is under review. Your podcast goes through a review process for suitability and appropriateness, which may take several days. When the review is completed, you are sent a further notification.

All music store customers can search for and subscribe to podcasts, as well as view podcasts on iTunes or download it to their iPod.

Podcasting your Skype personality to the world

Skype is a rich environment for podcast content. Some ambitious skypers are producing programs, which is a step up from casually recording a Skype conversation to help you recall the details of a meeting. You can create podcasts of interviews, poetry readings, performances, news reports, and conferences over Skype. Skypecasts, the large-scale audio conferences posted on the Skype Web site, can be captured, edited, and converted into podcasts as well.

Call Recorder, HotRecorder, Audio Hijack, SAM, Pamela, and Skylook can all be used to provide audio content for a podcast. However, some veterans of the podosphere prefer more rigorous recording equipment, and podcast studios can get pretty geeky. Getting truly clean audio is a tricky business, especially when your environment is virtual. A good book that covers scores of gadgets and techniques for producing podcasts is *Podcasting For Dummies*. It even has an overview of equipment and editing software to get the best audio production from Skype.

Part V
The Part of Tens

The 5th Wave   By Rich Tennant

" Dennis! Dennis!! Where are you?! I'm being attacked! I know, I know, but you're the groom. Tell them to just postpone the wedding a few minutes."

In this part . . .

We all like to make lists, and here are a few of ours. In this part, we offer ten ways to entice your mom and other family members to hang out together more often no matter how far apart you may be — and for free! We also describe ways to promote your business to the world with Web "presence" tools and local phone numbers on the other side of the globe. And finally, here are ten ways for teachers to use Skype-to-Skype calling for debates, book clubs, discussions, and research between classrooms across town or across the world.

Chapter 15

Ten Reasons Your Mom (and Other Family) Will Love Skype

· ·

*O*ur friend Jessica contacted us over Skype while we were writing this book. She lives in London, where Skype is wildly popular, and she joined in on the fun. When we asked her how she used Skype, she replied the same way virtually everyone responds: " I use it to talk to my mom."

Skype Is Free

The number one reason Mom would love Skype is that it is free. A free call through Skype means that you can stay on longer. It also means that Mom can leave the conversation to answer the door, baste the roast, or answer a business colleague's call without worrying about cost. With phones, time is money; words have a price on their head. Skype makes having long-distance conversations more like chatting in the same room.

Skype Is Easy

One of the authors of this book has a mom who is 84 years old. She hardly ever touches the computer except to look over her husband's shoulder (he's 92). But when Skype was installed and the startup was automated by having Remember My Password selected, Mom didn't hesitate to grab the mouse by its tail and click. The blue desktop icon with the big *S* in the middle is all Mom has to know to connect.

Relive Mom's Home Cooking from a Thousand Miles Away

Do special occasions always make you hungry for some mouth-watering childhood dish? (Perhaps you can still taste that strawberry shortcake your mom baked for your birthdays.) Sadly, you may have totally forgotten, mistakenly trashed, or never recorded Mom's recipes. You'd have to call Mom, crush the phone to your shoulder, and have her dictate the recipe (while spending as much money on the call as you would on the ingredients). Now, with Skype, Mom can tell you exactly how to make her Lemon Chicken, mix up her low-cholesterol Cocoa Cake, and much more while you're cooking. You can leave Skype on and ask questions (or get spontaneous tips). In the process, Mom can send you the recipe for good measure at the same time. Having a personal cooking coach makes life in the kitchen that much easier. Besides, Mom may use the "on air" practice to start a cooking show franchise on cable TV!

No-Sweat Party Planning

Organizing a family gathering can be a logistical nightmare, even if all the participants are local. Coordinating pickups, arrivals, and who is bringing the dessert can be handled by a conference call or conference chat. Even if Mom is not initiating the conference call, which requires more than one click (but not much more), she can still use this feature to connect with Paul in Newburgh, Leah in Flemington, and Susan in Brooklyn. If the plans are made with the message board, they persist, meaning that even after you quit the program, when you start up again the transcript of the last chat remains. You can even save all your planning chats with bookmarks (see Chapter 6). This is a very handy feature to help keep the planning straight if your pen runs out of ink or your calendar page runs out of writing room.

No-Sweat Messaging

Mom may or may not have been born to compute, but she's sure to want to take advantage of all its perks. A common family scenario is sharing travel plans. This used to be simple ("Mom, I'm coming in on Flight 123 on Blue Sky Airlines flying into YourCity terminal"). Now, parents, grandparents, and other interested parties want to know where you are every moment you're in the air. Doing so might require that you read a complicated URL or Web address over the phone so that family members can connect to a flight-tracking Web site. Dictating a URL with endless letters, numbers, underscores, slashes, and so on might require more time and patience than is at hand. The alternative, copying and pasting a URL into an email, may be equally frustrating because some email programs don't automatically translate Web addresses into live links. Using Skype messaging enables you to send Mom your itinerary with a live link to the airline to track your flight. If there is some confusion, you can simply talk about it and explain what to do on the spot. If Mom can hear it, and Mom can see it, Mom can do it (and that goes for the rest of us).

Keeping Track

Most of us lead busy lives, and moms are among the busiest. In Skype, messages are more than a passing thought. They can be kept indefinitely. Skypers can even go back in time. By choosing the History Tab in your main Skype window, you can see a long list of Skype activities, including past text chats. Click a chat item in your History list and you can reread the entire text conversation (see Chapter 6). In fact, you can even use the cute pushpin icon to bookmark a specific conversation. It's a great way for Mom to refresh her memory, refresh your memory, and keep track of details.

The History view looks like a script for a play, and you don't have subheadings and attachment files to wade through to get a snapshot of a conversation.

Guilt-Free Interruptions

Calling Mom or other family members at their jobs is a bit of a nail-biter for many people. Using a landline or cell phone, you never really know whether you are calling at a bad time. On Skype, the status icons that announce your availability alleviate the worry.

Chapter 3 tells you all about setting your online status icons.

Guilt-Free Reminders

Using the chat window as a just-in-time memory jogger in the background is a great way for Mom to be in contact when she needs to let you in on some news. The neat thing about Skype is that even though you or your mom may have posted the Offline status icon, you can still send each other a message. On the Mac platform, Skype will even let each of you know, ever so politely, that you "may not respond."

Doing so doesn't stop the message from getting through. This method sure beats writing down that fleeting thought on a scrap of paper that disappears among all the other scraps of paper and leads to the "There was something I wanted to tell you" conversation. Mom can simply send a message before it slips her mind (and vice versa), even while she's waiting to board a plane for a business trip. Juggling families and careers can be made just that much easier with Skype.

Stamp-Free Announcements

Although receiving an invitation in the mail is lovely, there are just some events that most of us can't wait to tell the world. Besides, not all announcements are "stamp worthy"; you're not likely to send an "I bowled 300!" card. Using "Mood Messages" (see Chapter 3), you can add a comment that appears next to your Skype Name and is visible to all your contacts. It's like having a virtual public address system. Although Skype calls this a Mood Message, meant to announce your state of mind to the world, there's nothing to stop you or Mom from using it for more practical things. She can post a "hold the date" message for upcoming celebrations, let everyone know her travel schedule, or even announce, "'I'm running the Boston Marathon on Sunday." The possibilities are endless!

Keep in Touch and Stay in Sight

With Skype, busy moms can see as well as hear their children, which is especially nice if their work takes them out of town. It is probably one of the best reasons for parents, who often live far away from grown children, to use Skype. Setup is easy and the payoff is huge. Moms can see how their children are doing. Older moms can visit with grown children and see their grandchildren. Younger moms may have school-age children at home with a nanny while they're at work. The video camera lets a mom traveling on business see whether homework is done and school forms are filled out. Video contact gives mom a feeling of well-being because she can check in — and see for herself — that everything is okay.

Families need a common arena. They may be scattered many miles or countries apart. Discovering a free forum that enables a very flexible method of communicating (voice, written messages, video, announcements) can make it easier for families to get closer. Phone calls lead to phone tag. Emails are handy, but they are limited and out of sync with a casual conversation. Skype helps families to see, hear, write, and archive messages with one another at no cost beyond their Internet service bill.

Chapter 16

(Almost) Ten Ways to Promote Your Business Using Skype

* *

*H*ere are ten cost-free ways to hypercharge your business and make it both global and personal at the same time. Promoting your business is not limited to an ad in the paper.

Notify Your Customers of Important News

Businesses can easily spend between 5 to 30 percent of their budget on advertising. For a small business, advertising is a make-or-break expenditure. For a large business, the choice is to spend the money or don't be large anymore. Skype has a sweet little utility that can grab the attention of customers at the cost of, well, nothing. Called *Mood Messages*, these are comments that you can post beside your Skype Name in a Contact list. They were originally designed to indicate your mood to the world ("I'm on top of the world"; "I'm in the zone today"; "The world is my oyster") so that your contacts would know your state of mind before calling.

For businesses, this feature is like having a PA system broadcasting to the world. A business can use these messages to announce events, special sales, and press releases, and the announcement appears in the Contacts window of every person on your Contacts list. A subtle benefit is that your customers feel more closely connected to your company even when you don't have the time and bandwidth to reach everyone personally.

Here are some ideas for Mood Messages to promote your business:

- Fifty percent off sale starts today!
- Earnings and profits tripled over the last year!
- A shipment of 20 million vaccine doses delivered.

> ✔ We're now open Thursday evenings till 10:00.
>
> ✔ Register your birthday and get a free dinner!
>
> ✔ Stockholders approve merger.

You can create a Mood Message in the Skype main window. Click your Skype Name (see Figure 16-1) and enter your message.

Figure 16-1:
Setting your
Mood
Message.

Use SkypeWeb Alerts to Get Customers

What do sales people, actors, models, photojournalists, babysitters, and dog groomers have in common? They all need to announce their availability to customers. Using the SkypeWeb alert lets the world know you are ready for business the minute you log on to Skype and choose your Skype Me option (see Chapter 2). The SkypeWeb alert is a status button that signals your availability and a direct contact link to you. You can post a SkypeWeb alert on your home page, a blog, an email signature block, and basically any Web page that gives you an opportunity to post contact information. The SkypeWeb button not only automates this process but also puts you literally one click away from anyone who needs to reach you from the Web. See Chapter 12 for details about getting and using the SkypeWeb button.

Make Your Business "Local" Anywhere

You live in Kansas, busily manufacturing ruby slippers that have suddenly become all the rage in London and Sydney as well as in your hometown. You want to build up the buzz around your product and make it easy for your customers to purchase the popular ruby slippers with a local phone call. After all, a great portion of sales comes from people buying either out of convenience or on impulse.

In a jiffy, you can establish SkypeIn phone numbers (see Chapter 8) that give you a local presence in London, or Sydney, in addition to your physical presence in Kansas — and you can do it for less than $100 per year. Customers now have a local and convenient phone number for product orders and customer service, with customers from each continent on a different time zone.

Businesses that respond to the market quickly by promoting their products can more ably compete, survive, and flourish. When you can establish a local business presence across multiple continents for less than $100, you know you're not in Kansas anymore!

Conduct a Global Town Hall Meeting

A powerful way to promote your business is by simultaneously promoting a good relationship with the community you serve. Skypecasting (see Chapter 14) provides a convenient and inexpensive venue for accomplishing these goals. A local health clinic, for instance, may Skypecast an information session on controlling cholesterol or providing support information for cancer survivors and their family.

The community feels connected to that institution or company because they are able to get quality information and answer questions directly. This kind of Skypecast engenders trust, and trust translates to a larger customer base.

Informational Skypecasts can work for real estate firms explaining tax abatement programs; pension consultants providing workshops on retirement benefits and strategies; law firms providing guidance on estate planning; accounting firms explaining the implications of new financial reporting and disclosure procedure; product safety associations providing consumer information; and financial consultants and accountants offering valuable tips on how you can protect yourself from fraud and identity theft. In short, just about any business with a need to provide consumer information outside the sales cycle can greatly benefit from Skypecasting.

Mentoring and Training

Visiting your client's office for one-to-one personal training and mentoring helps to build a healthy customer relationship. Having that physical presence, especially when that client is a 1,000-plus miles away, is costly for both you and the client. Why travel? The advent of video conferencing, Web-based collaboration, and the ability to have both you and the client interactively taking turns driving the mouse and keyboard, result in an alluring medium for one-on-one mentoring, all with zero travel time and cost.

Here's how to do it. Start with some Web conferencing software (see the links on our Web site at www.skype4dummies.com). Set up your client on Skype. Call your client over Skype and post a link to your private Web conferencing and mentoring session. The client clicks the link, and within minutes, both of you are actively engaged working together on a spreadsheet, a requirements document, or whatever else fits the occasion. You can stop the session and resume at any time, transfer files, post links on a Skype chat, and add others to the Web mentoring session. This venue is sure to be a hit with your clients. (See Chapter 10 for more about Web conferencing.)

Skype on a Business Card

These days, a business card typically contains your phone, email address, and perhaps a Web site. So, why add your Skype ID? Well, during the late 1980s, people were beginning to place their email address on their business cards, even though it raised a few eyebrows. In 1995, people started including their Web address on their business cards before it became popular. Now it's commonplace — in fact, often required — to add email addresses and Home page URLs to contact information. No one questions why. Soon, a Skype Name may be as ubiquitous.

People value the ability to see around corners for that next big thing. Placing your Skype Name signals that you're onto something — and, more important, somewhere during your career, you're likely to find that having that Skype connection will land you a contract, engagement, or a new job.

Large-Scale Online Conferencing

Imagine trying to set up an audio conference call to accommodate the 500 or so shareholders in your public company as you review quarterly earnings. If you provide a toll-free number, the call can cost you something like ten cents a minute per caller. This amounts to $50 per minute, or $3,000 per hour. Ouch!

An alternative approach is to use a Skype-related service that allows Skype users to dial in from their Skype software to the audio conference at no cost. This free service can accommodate up to 500 attendees at a time.

To take advantage of this service, sign up for a free account at (www.highspeedconferencing.com) and begin using your dedicated conference calling number right away (see Chapter 12).

Promotionals and Giveaways

We've all come across a pen that has a company name and logo inscribed in gold or silver lettering. This sort of advertising is a large industry that manufactures thousands of give-away items such as address books, balloons, business card holders, bags or totes, and many more. Rather than give your customers a free calendar or tote bag, how about two hours of SkypeOut credit? Who wouldn't love to have a slick speakerphone with loads of SkypeOut minutes included in the purchase price? Your customers will rave about the benefits of buying from your business, and this may pave the way for some well-deserved sales.

Promotionals with Skype are bound to expand. It may be only a matter of time before the airlines are trading travel miles for Skype minutes. The same is true with credit card companies for purchases, and banks for opening accounts.

Improve Customer Service with Skype Call Transfer

If you're a business user seeking to set up call center functionality or a software developer creating Customer Relationship Management–like applications, then you're in luck. Skype can provide you with the means of dynamically routing and transferring incoming calls from one Skype user to another. Here's one scenario: Your customer makes a free call using Skype. The call center recognizes the Skype Name of the customer and routes the call to the appropriate party. No need for an operator or extension numbers. Instead of passing the buck, you're passing the benefits of decreased customer servicing expenses with higher satisfaction — a win-win combination.

Chapter 17

Ten Ways to Use Skype at School

Schools have outgrown inkwells, slates, and typewriters. Email, instant messaging, and the Internet have opened the classroom to the world. Skype empowers students and teachers even further by adding features for collaboration, live video, and instant file sharing. This chapter offers some ways to energize school and learning with Skype.

Connect to the World on a Teacher's Budget

Telephones are a scarce commodity in classrooms. Telephones with international calling plans . . . well, we haven't seen one of those yet in a classroom (unless it's a teacher's personal cell phone). The impact and benefit of students speaking with students across the globe brings an authenticity to education that adds immeasurably to both textbooks and Web references. Skype makes it easy to connect to other people, other cultures, and other countries. Classrooms may not have telephones, but they are often wired for the Internet. Skype is a service that teachers and students can access using the hardware they already have in their rooms. It's common knowledge that teachers spend, on average, $500 out of their own pockets to stock their classrooms. Skype fits into a teacher's budget. It's free.

Master a Foreign Language (Or Practice a Phrase)

The old pen-pal model takes on a new approach when the pen is a mouse and a microphone and the paper is virtual. Having students speak to each other over Skype benefits each language group simultaneously. Of course, the knotty question is, how do you find safe, viable, dynamic forums for second-language practice?

A Web-savvy teacher with a desktop and a high-speed line can create a gathering place to reach language learners. Schools from Italy, the United States, Turkey, Sweden, England, and more have connected with each other over Skype to help students master foreign languages through educator-mediated Web forums (see Chapter 13). The schools control their projects and connectivity. A main Web site functions like a town square, a place to find kindred spirits and post Skype contact names. Schools sign up educators, contact each other, and arrange student conversations. Teachers are in charge and students have a blast.

Have School Beyond the Classroom Door

Ten years ago, one of the authors (Susan) attended an educational conference in London. One of the presenters, a teacher from Buffalo, New York, revealed that although she was in London, her English writing class was still on schedule. At the time, Susan was astonished to discover a use for the brand-new technology, email, in education. Students couldn't skip their writing assignments and couldn't blame the dog, the spilled soup, or the late bus. Their teacher, on the other hand, couldn't even suspend class because she was across the ocean.

That teacher-across-the-ocean model is even easier with Skype. Students can still transfer over their assignments, but not as attachments that may be garbled by email protocols, or as pasted text that loses all its formatting. A quick file transfer, in mid-chat or mid-conversation, brings the assignment from student to teacher in record time. In addition, teachers can use the Mood Message option to post a note next to his or her Skype Name to let students know what topic is on the agenda, when an assignment is due, or just that they are doing great work. Remote teachers can encourage students with similar interests to work together, encourage older students to mentor younger ones, or simply hold class with the Skype conference facility.

Provide Professional Development

Skype brings professional development to the teacher, coach, or mentor when and where they need it. The common model of improving teaching skills gathers teachers in one place for a one-size-fits-all workshop. With Skype, teacher mentors can deliver personalized training directly to the classroom on subjects teachers need. Whether the interaction occurs one on one, through a mini-conference, or on an as-needed advice line, using Skype skips the expense, the travel, and the time out of the classroom for teachers.

Here's a real-world example: A staff developer in the New York City school system is charged with supporting the instructional technology for students who are blind. The teachers he works with often have questions about voice output technology and other assistive software installed on students' desktop computers. The Skype environment provides needed guidance "on-site," on the very computer that students need to use (and that teachers need to configure).

Encourage Student Collaboration

Skype is a natural environment for student collaboration. In one software setup, students can talk with each other and send drafts of files to everyone in the group to review and interject with text comments. Archives of Skype text messaging conversations can bring fellow students up-to-date if they miss a meeting. Students can work on school newspapers, use video to practice a senior play, try debate strategies, and even organize school events.

Host Poetry Slams, Debates, and Book Clubs

Organizing a competition, performance, or club involves much decision-making beyond picking a time and place to host these activities. But if the time and the place were virtual, most of the effort would involve the educational content of the activity rather than endless logistical planning. Children in different schools, counties, and even countries could gather at arranged times to discuss books, deliver a book report, or read their own stories. A Skype voice and videoconference can, and does, serve as an exciting international locale for poetry performance or "slams" (competitions). Each class that participates projects the Skype conference and video windows with an LCD projector so that everyone can see the action.

Five to ten classes (or venues) can conference an event. Student poets or debaters can face their counterparts anywhere in the state, the country, or even the world. The students can rate one another's performance, offer comments, expand viewpoints, or simply post an emoticon in a simultaneous chat window to provide another layer of participation.

In an increasingly virtual environment, it is not just fun but also necessary for students to hone presentation skills for a virtual audience. Making a speech is scary. We all know what it's like (butterflies in the stomach, sweaty palms, weak knees) to face a class and deliver a research paper or argument (or even make a toast at a wedding). The same dread happens with online presentations, and developing an online comfort level when facing virtual audiences is an important skill. The opportunity to start practicing in a safe place — the classroom — and in a cost-free environment — Skype — is a great beginning.

Record a Group Thought Process

Students get together to prepare material, study for exams, or work on after-school club activities. But sometimes a student who is busy hatching a great idea may not be listening to the other great ideas being put forth by his or her student collaborators. Here's where voice recording comes in handy: You can use third-party software to make a sound file of a Skype call or conference. Play it back to hear all the nuances you missed or to find all the facts you didn't have time to jot down (see Chapters 9, 11, and 14). Record a dramatic dialogue, foreign language pronunciation, debate arguments, or an oral presentation as it's happening for instant feedback to provide students with help on making improvements.

Use Skype as a Homework Helper

Although some large cities have extensive Web sites with live homework helpers ready to guide a youngster through a difficult assignment, smaller towns may not have the means to set up an after-school resource center. Skype can double as a quick connection to a teacher, librarian, or even a traveling parent for a child in the midst of a homework crisis.

Conduct Read-Alouds

One of the most important ways to inspire children to read is by reading aloud to them. Schools can arrange to have an author read a story over Skype so that the entire class can enjoy a favorite book and then ask questions for an author study. Other remote readers might be local business owners, elected officials, athletes, or other local leaders. Some local theater groups get involved in literacy education by offering read-alouds in classrooms. They can expand their service by adding a Skype call to the classroom read-aloud. It would be like the old radio shows that used to broadcast to the world while performing in front of a live audience.

But Skype storytelling is not limited to school hours or even schools. A nightly Skype story delivered by a working or traveling parent can continue a valuable routine that job responsibilities might interrupt. A creative grandmother can read stories to all her grandchildren at one time in an organized conference.

Support Special-Needs Students

Children with special needs often have unique problems that make the school experience difficult. Some children with developmental delays, autism, or speech and language problems cannot easily describe their day when they return home after school. Teachers write notes, phone when they can, and try to make up for the lack of communication during open school night. With Skype, teachers can send a description of the day over a chat or file transfer, and offer Skype conference hours after school or during administrative periods during the school day. Teachers can also help direct a parent with a struggling learner by modeling teaching techniques. The parent and child are at home and the teacher coaches through Skype. The natural setting and the convenient, supportive, communication change the very nature of the parent/teacher conference!

Skype can also offer a unique combination of accessible modes of language. Children who are blind can join in an audio Skype conference. Children who are hearing impaired can engage in a written chat. Children who use sign language can use the video mode as well as the chat for maximum flexibility. Children who are homebound can actually join in a school-based classroom without expensive teleconferencing equipment. Skype connects the special-needs student with new opportunities for learning.

Appendix A

Skype Multilanguage Support

Do you speak French, Swedish, or Turkish and want to set your Skype window menus in those languages? Skype supports 27 different languages and enables you not only to load any one of them but also edit individual menu items. You can add an entirely new language to your Skype window, or just make small changes.

Changing Skype to Your Language of Choice

Setting your language of choice is easy. With the Skype main window open, choose Tools⇨Change Language and pick a language from the list that appears (see Figure A-1). Your Skype menus are instantly changed.

You can easily edit and alter all the menu items and text labels that appear in Skype to suit your own needs. Here's how:

1. **Choose Tools⇨Change Language⇨Edit Skype Language File.**

   You see a table with 2,600 editable Skype menu items (see Figure A-2).

2. **To edit an item, select it by clicking in a field in the Current column.**

   For example, click the status alert Do Not Disturb. Change the text a little to Do Not Disturb Please. Click Apply so that Skype knows to apply your changes. Repeat this step for every Skype term you want to edit.

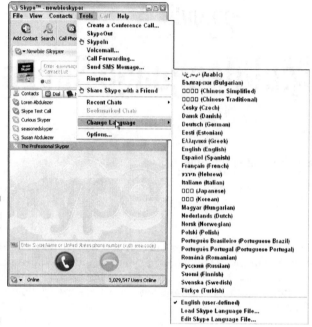

Figure A-1:
Choose your
language
from among
Skype's 27
Language
options.

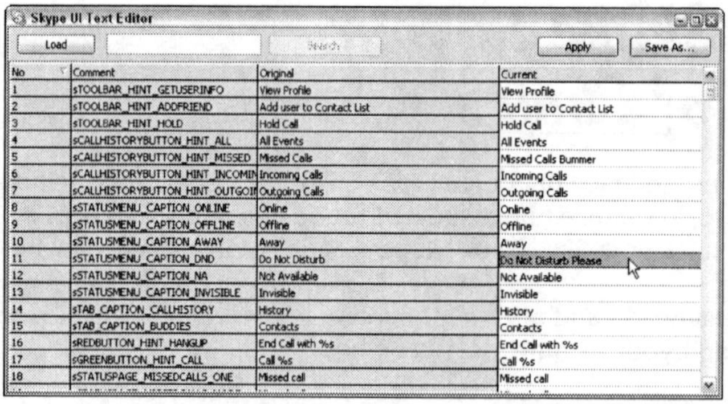

Figure A-2:
The Skype
language
editing
table.

3. **Click Save As and adjust the language filename, which is saved as a .lang file (see Figure A-3).**

   Saving your customized language file allows you to load these applied changes to other computers on which you may be using Skype. This file can be emailed, archived to a CD-ROM or flash drive, or even transferred to other Skype users through Skype's file transfer facility (see Chapter 6).

Figure A-3:
Saving a
customized
language
file.

If you followed this example, you can verify that the availability menu is
customized. To do so, click the status icon on the lower-left corner of your
Skype window (see Figure A-4). It's changed!

Figure A-4:
Menu with
edited
phrase.

When you are using Skype on a different computer and want to load your customized Skype language file, choose Tools➪Change Language➪Load Skype Language File.

Don't miss the Skype Translation Forum located at:

`http://forum.skype.com/index.php?showforum=7`

You may find additional language files that weren't bundled with your Skype application.

Appendix B

Skype Tips and Tricks Guide

In This Appendix

▶ Troubleshooting some common issues

Troubleshooting Skype

*M*uch of the popularity of Skype popularity is owed to the fact that Skype is easy to use. Even though Skype is designed for simplicity, there are bound to be situations in which a complication or two arises. Here are some questions and answers that may help smooth the way through unexpected roadblocks.

Table B-1	Troubleshooting Common Issues with Skype
What's the Problem?	*What's the Answer?*
My Contacts search yields too many names and I can't find the person I'm looking for.	It helps to sort the list by Country/Region. Just click the column header of the search results. Entering more precise search terms also helps. The more data you enter, the better the search results.
I signed up with Skype, but no one can find my Skype Name.	Add more information to your profile to yield better search results for your new contacts. At the very least, fill in your country and language.
I want a secure password, but it needs to be one I can remember.	Pick two easy but unrelated, words. Separate them by a number, such as latte67chair.
I can't find the audio jack in my Macintosh computer. How do I use a headset/ microphone with it?	Macintosh accepts audio input from USB devices. Purchase a USB microphone or headset.

(continued)

Table B-1 *(continued)*

What's the Problem?	What's the Answer?
I misplaced my microphone, and no one I speak with using Skype can hear me.	In a pinch, use an extra set of headsets that you plug into the microphone jack. Not the best quality sound, but it works.
I put my photo in my profile, but now it's gone.	Profile information is stored in a central Skype database except for photos. You will have to upload your photo to your profile from each local machine running Skype.
I have 50 contacts, but I want to chat with only 15. That's a lot of scrolling, clicking, and inviting.	Create separate groups in your Contacts list (family, club, softball buddies) and add contacts from your Skype Contacts list to your new group (see Chapter 3). Choose View⇨Show Contact Groups. Click the contact group of your choice. Press Ctrl+A to select all the contacts in the group and start the chat as you normally would (see Chapter 6).
I want to transfer a file to 15 chat buddies simultaneously. What do I do?	Choose View⇨Show Contact Groups to view the various groups you have created. Right-click the group you want to send a file to and select Send File to Group. You will need to locate and open the file you want to send. A quicker way to send the file is to just drag the file onto the group.
I've changed the font style in my chat window and I want to go back to the default but don't remember what font it is.	Tahoma, 8 points, Western Script.
I have too many devices attached to my computer during a video conference. Help!	Get a webcam with a built-in microphone and noise cancellation. Your life will get easier!
I just installed a video driver, but my video is not working correctly.	Make sure that you turned off anti-virus and anti-adware software before installing video drivers.
I just installed a new webcam, but it's not working properly.	Make sure that you plug in the webcam when the software prompts you. Most video drivers are installed first; the webcams are plugged in after installation.
My video is not working properly.	Check for the latest updates from your webcam's Web site.

What's the Problem?	What's the Answer?
I put my SkypeIn number on my business card and now I'm getting calls at 2 o'clock in the morning!	Always indicate your local time zone when giving out your SkypeIn number.
I don't hear anything through my USB headsets.	Plug in your headsets before your start Skype. If you've started Skype, quit and restart.
I have a Bluetooth headset to use with my Macintosh speech-recognition software, but nothing is happening.	A Bluetooth headset does not work with Macintosh speech recognition. However, you can use a Plantronics CS50-USB wireless headset (it's not Bluetooth).
I'm trying to SkypeOut, but it's not working.	Remember to enter a plus sign (+) before a country code and then enter the telephone number.
I set up a Skype Internet Telephone with a USB base station. Now I don't hear the music from my computer's MP3 player.	Check your audio settings. Some Internet phones become the default audio device and work only with Skype.
I'm using my Wi-Fi mobile device to make a Skype call. I want to see my contact's Skype profile during the call; what do I do?	During a call, you can tap your mobile screen twice to bring up big call function buttons. Click the Profile button for info.
I'm using my Wi-Fi mobile device to make a Skype call and want to mute my voice during the call.	During a call, you can tap your mobile screen twice to bring up big call function buttons. Click the Mute button.
I'm using my Wi-Fi mobile device to make a Skype call and want to add the caller to my Contacts list.	During a call, you can tap your mobile screen twice to bring up big call function buttons. Click the Add Contacts button.
Skylook records my conversations. Where do I find them?	Skylook creates a Conversations folder in your Outlook inbox.
Skylook records my conversations, but I don't want it to do this anymore.	Click the orange Skylook options icon in your Skylook toolbar. Click Options and then click the Recording tab. Uncheck the Record Voice Call Audio check box.

(continued)

Table B-1 *(continued)*

What's the Problem?	What's the Answer?
I'm using Pamela. I want to record sometimes, but not all the time.	Ignore the recording prompt. It will go away after 10 seconds and not record anything.
When I'm in a Skype conference and transferring a file, our voices get harder to understand.	Put the conference on hold, transfer your files, and then resume the conference. The files will transfer faster and your voice quality won't suffer.
I'm on the road a lot. How do I skype when I'm away from my wireless network?	You can subscribe to a service, such as Skypezones, that has thousands of wireless access nodes to pick from as you travel.
I'm feeling adventurous and want to do a Skypecast. What do I do?	Sign up to offer a Skypecast at `https://skypecasts.skype.com/skypecasts/home` and read Chapter 14 of this book.
How do I let skypers know what time zone I'm in?	Include your time zone in your Skype Profile. If you don't know your time zone, see the cheat sheet in the front of this book.
I want to find out my Skype contact's email address; how do I search for it?	You can't through Skype; email addresses are not revealed. Just ask your contact.
I have a webcam that I use with Skype, but I'm afraid someone will see me in my pajamas.	Don't worry. Skype video does not start automatically. You have to select Start My Video to activate it.
Can I make an emergency call on Skype?	No. You must use a regular phone.
I want to purchase SkypeIn and SkypeOut. What do I need to do this?	You must have a Skype password and a valid email address to purchase Skype credit.
I have Skype credit. Do I need to use it up in the 180-day time limit?	No. Just keep it alive by making one call within 180 days to extend it for another 180 days. If you don't use it at all, however, it will be lost after 180 days.
I have Outlook Express. Will Skylook work?	No. Skylook works only with Outlook.

What's the Problem?	*What's the Answer?*
I have Skylook and can see my Skype contacts in Outlook. Can I see my Outlook contacts in Skype?	Yes. In the Skype menu, choose View⇨View Outlook Contacts.
I have a huge list of Outlook Contacts showing in my Skype Contacts list. What do I do?	This is happening because you have a check mark next to View Outlook Contacts in your Skype View menu, To remove the check mark, choose View⇨View Outlook Contacts.
I want people to use regular telephone directory assistance to get my Skype information. Is this possible?	Yes. If you sign up for services such as listyourself.net, anyone can find out your Skype Name through an ordinary directory assistance telephone call. Your Skype Name is distributed to and entered into regular databases used by public information call centers.
My Skype Profile photo is not the right size. What size should it be?	Profile photos should be 96 by 96 pixels.
I have a poor Skype connection. What can I do to make it clearer?	Hang up. Set your Skype status to Offline for 15 seconds and then call again. With luck, the connection is reset and the sound quality is higher.
I have a long Contacts list and I'm afraid I will lose the information if my computer crashes. How do I back my list up?	You will find your contacts in the following folder on your hard drive:C:\Documents and Settings*yourusername*\Application Data\Skype and you can back up that folder on a USB Smart Drive or a CD-ROM. Because the file is small, you can even save it to a floppy disk or email it to yourself for safekeeping.
I'm trying to transfer a file to my friend Bob, and Skype won't let me, even though I've done this many times before. What can I do?	Have Bob take you off his Contacts list. Make contact again and agree to share contacts. Your files should go through smoothly now.
I forgot my Skype Name!	Enter the following in your Internet Explorer browser's URL address field: **%AppData%\Skype** You will see a folder labeled with your Skype Name. If you have more than one Skype Name, you will see any others, too.

(continued)

Table B-1 *(continued)*

What's the Problem?	What's the Answer?
I've plugged in my USB headset, and now I can't hear Skype ringing unless my headset is on. What can I do?	On the Skype menu, choose Skype Tools⇨ Options⇨Sound Devices. The Audio In and Audio Out options should be set to Windows default audio or to your USB headsets. Also, check the Ring PC Speaker check box.
I want my Skype status icon to remain on Available. It keeps changing even though my Skype program is running.	On the Skype menu, choose Tools⇨Options⇨ General and set the Show Me as Away When I Am Inactive option to zero. Do the same for Show Me as Not Available When I Am Inactive.
I forgot my Skype password.	Go to www.skype.com/go/ forgotpassword and fill in the form.
I want to be a slick skyper.	Read *Skype For Dummies*. All of it.

Index

• *G* •

• *H* •

• I •

• J •

• K •

• *X* •

USINESS, CAREERS & PERSONAL FINANCE

0-7645-9847-3

0-7645-2431-3

Also available:
- Business Plans Kit For Dummies
  0-7645-9794-9
- Economics For Dummies
  0-7645-5726-2
- Grant Writing For Dummies
  0-7645-8416-2
- Home Buying For Dummies
  0-7645-5331-3
- Managing For Dummies
  0-7645-1771-6
- Marketing For Dummies
  0-7645-5600-2

- Personal Finance For Dummies
  0-7645-2590-5*
- Resumes For Dummies
  0-7645-5471-9
- Selling For Dummies
  0-7645-5363-1
- Six Sigma For Dummies
  0-7645-6798-5
- Small Business Kit For Dummies
  0-7645-5984-2
- Starting an eBay Business For Dummies
  0-7645-6924-4
- Your Dream Career For Dummies
  0-7645-9795-7

OME & BUSINESS COMPUTER BASICS

0-470-05432-8

0-471-75421-8

Also available:
- Cleaning Windows Vista For Dummies
  0-471-78293-9
- Excel 2007 For Dummies
  0-470-03737-7
- Mac OS X Tiger For Dummies
  0-7645-7675-5
- MacBook For Dummies
  0-470-04859-X
- Macs For Dummies
  0-470-04849-2
- Office 2007 For Dummies
  0-470-00923-3

- Outlook 2007 For Dummies
  0-470-03830-6
- PCs For Dummies
  0-7645-8958-X
- Salesforce.com For Dummies
  0-470-04893-X
- Upgrading & Fixing Laptops For Dummies
  0-7645-8959-8
- Word 2007 For Dummies
  0-470-03658-3
- Quicken 2007 For Dummies
  0-470-04600-7

OOD, HOME, GARDEN, HOBBIES, MUSIC & PETS

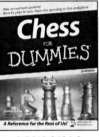

0-7645-8404-9

0-7645-9904-6

Also available:
- Candy Making For Dummies
  0-7645-9734-5
- Card Games For Dummies
  0-7645-9910-0
- Crocheting For Dummies
  0-7645-4151-X
- Dog Training For Dummies
  0-7645-8418-9
- Healthy Carb Cookbook For Dummies
  0-7645-8476-6
- Home Maintenance For Dummies
  0-7645-5215-5

- Horses For Dummies
  0-7645-9797-3
- Jewelry Making & Beading For Dummies
  0-7645-2571-9
- Orchids For Dummies
  0-7645-6759-4
- Puppies For Dummies
  0-7645-5255-4
- Rock Guitar For Dummies
  0-7645-5356-9
- Sewing For Dummies
  0-7645-6847-7
- Singing For Dummies
  0-7645-2475-5

NTERNET & DIGITAL MEDIA

0-470-04529-9

0-470-04894-8

Also available:
- Blogging For Dummies
  0-471-77084-1
- Digital Photography For Dummies
  0-7645-9802-3
- Digital Photography All-in-One Desk Reference For Dummies
  0-470-03743-1
- Digital SLR Cameras and Photography For Dummies
  0-7645-9803-1
- eBay Business All-in-One Desk Reference For Dummies
  0-7645-8438-3
- HDTV For Dummies
  0-470-09673-X

- Home Entertainment PCs For Dummies
  0-470-05523-5
- MySpace For Dummies
  0-470-09529-6
- Search Engine Optimization For Dummies
  0-471-97998-8
- Skype For Dummies
  0-470-04891-3
- The Internet For Dummies
  0-7645-8996-2
- Wiring Your Digital Home For Dummies
  0-471-91830-X

Separate Canadian edition also available
Separate U.K. edition also available

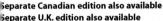

ailable wherever books are sold. For more information or to order direct: U.S. customers visit www.dummies.com or call 1-877-762-2974.
K. customers visit www.wileyeurope.com or call 0800 243407. Canadian customers visit www.wiley.ca or call 1-800-567-4797.

SPORTS, FITNESS, PARENTING, RELIGION & SPIRITUALITY

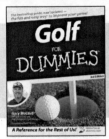
0-471-76871-5

0-7645-7841-3

Also available:
- Catholicism For Dummies
  0-7645-5391-7
- Exercise Balls For Dummies
  0-7645-5623-1
- Fitness For Dummies
  0-7645-7851-0
- Football For Dummies
  0-7645-3936-1
- Judaism For Dummies
  0-7645-5299-6
- Potty Training For Dummies
  0-7645-5417-4
- Buddhism For Dummies
  0-7645-5359-3

- Pregnancy For Dummies
  0-7645-4483-7 †
- Ten Minute Tone-Ups For Dummies
  0-7645-7207-5
- NASCAR For Dummies
  0-7645-7681-X
- Religion For Dummies
  0-7645-5264-3
- Soccer For Dummies
  0-7645-5229-5
- Women in the Bible For Dummies
  0-7645-8475-8

TRAVEL

0-7645-7749-2

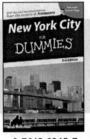
0-7645-6945-7

Also available:
- Alaska For Dummies
  0-7645-7746-8
- Cruise Vacations For Dummies
  0-7645-6941-4
- England For Dummies
  0-7645-4276-1
- Europe For Dummies
  0-7645-7529-5
- Germany For Dummies
  0-7645-7823-5
- Hawaii For Dummies
  0-7645-7402-7

- Italy For Dummies
  0-7645-7386-1
- Las Vegas For Dummies
  0-7645-7382-9
- London For Dummies
  0-7645-4277-X
- Paris For Dummies
  0-7645-7630-5
- RV Vacations For Dummies
  0-7645-4442-X
- Walt Disney World & Orlando
  For Dummies
  0-7645-9660-8

GRAPHICS, DESIGN & WEB DEVELOPMENT

0-7645-8815-X

0-7645-9571-7

Also available:
- 3D Game Animation For Dummies
  0-7645-8789-7
- AutoCAD 2006 For Dummies
  0-7645-8925-3
- Building a Web Site For Dummies
  0-7645-7144-3
- Creating Web Pages For Dummies
  0-470-08030-2
- Creating Web Pages All-in-One Desk
  Reference For Dummies
  0-7645-4345-8
- Dreamweaver 8 For Dummies
  0-7645-9649-7

- InDesign CS2 For Dummies
  0-7645-9572-5
- Macromedia Flash 8 For Dummies
  0-7645-9691-8
- Photoshop CS2 and Digital
  Photography For Dummies
  0-7645-9580-6
- Photoshop Elements 4 For Dummies
  0-471-77483-9
- Syndicating Web Sites with RSS Feed
  For Dummies
  0-7645-8848-6
- Yahoo! SiteBuilder For Dummies
  0-7645-9800-7

NETWORKING, SECURITY, PROGRAMMING & DATABASES

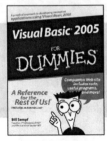
0-7645-7728-X

0-471-74940-0

Also available:
- Access 2007 For Dummies
  0-470-04612-0
- ASP.NET 2 For Dummies
  0-7645-7907-X
- C# 2005 For Dummies
  0-7645-9704-3
- Hacking For Dummies
  0-470-05235-X
- Hacking Wireless Networks
  For Dummies
  0-7645-9730-2
- Java For Dummies
  0-470-08716-1

- Microsoft SQL Server 2005 For Dummie
  0-7645-7755-7
- Networking All-in-One Desk Referenc
  For Dummies
  0-7645-9939-9
- Preventing Identity Theft For Dummies
  0-7645-7336-5
- Telecom For Dummies
  0-7645-7325-7
  0-471-77085-X
- Visual Studio 2005 All-in-One Desk
  Reference For Dummies
  0-7645-9775-2
- XML For Dummies
  0-7645-8845-1

HEALTH & SELF-HELP

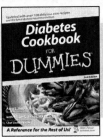

0-7645-8450-2

0-7645-4149-8

Also available:

- Bipolar Disorder For Dummies
  0-7645-8451-0
- Chemotherapy and Radiation
  For Dummies
  0-7645-7832-4
- Controlling Cholesterol For Dummies
  0-7645-5440-9
- Diabetes For Dummies
  0-7645-6820-5* †
- Divorce For Dummies
  0-7645-8417-0 †

- Fibromyalgia For Dummies
  0-7645-5441-7
- Low-Calorie Dieting For Dummies
  0-7645-9905-4
- Meditation For Dummies
  0-471-77774-9
- Osteoporosis For Dummies
  0-7645-7621-6
- Overcoming Anxiety For Dummies
  0-7645-5447-6
- Reiki For Dummies
  0-7645-9907-0
- Stress Management For Dummies
  0-7645-5144-2

EDUCATION, HISTORY, REFERENCE & TEST PREPARATION

0-7645-8381-6

0-7645-9554-7

Also available:

- The ACT For Dummies
  0-7645-9652-7
- Algebra For Dummies
  0-7645-5325-9
- Algebra Workbook For Dummies
  0-7645-8467-7
- Astronomy For Dummies
  0-7645-8465-0
- Calculus For Dummies
  0-7645-2498-4
- Chemistry For Dummies
  0-7645-5430-1
- Forensics For Dummies
  0-7645-5580-4

- Freemasons For Dummies
  0-7645-9796-5
- French For Dummies
  0-7645-5193-0
- Geometry For Dummies
  0-7645-5324-0
- Organic Chemistry I For Dummies
  0-7645-6902-3
- The SAT I For Dummies
  0-7645-7193-1
- Spanish For Dummies
  0-7645-5194-9
- Statistics For Dummies
  0-7645-5423-9

Get smart @ dummies.com®

- **Find a full list of Dummies titles**
- **Look into loads of FREE on-site articles**
- **Sign up for FREE eTips e-mailed to you weekly**
- **See what other products carry the Dummies name**
- **Shop directly from the Dummies bookstore**
- **Enter to win new prizes every month!**

* Separate Canadian edition also available
† Separate U.K. edition also available

Available wherever books are sold. For more information or to order direct: U.S. customers visit www.dummies.com or call 1-877-762-2974.
U.K. customers visit www.wileyeurope.com or call 0800 243407. Canadian customers visit www.wiley.ca or call 1-800-567-4797.

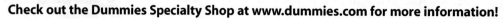

Make Skype fun!

Pamela for Skype is your own personal digital assistant for Skype which makes Skype even more effective and fun! Pamela is a Skype-certified software add-on that enhances your Skype with a lot of cool functions, including

- **Skype Answering Machine**
- **Auto Chat Replies**
- **Skype Call/Video/Chat Recording**
- **VideoMail (Video Answering Machine)**
- **Email Forwarding**
- **Personalization of Skype Contacts**
- **Birthday Reminders**
- **Skype Voice Mail Support**

Get 20% off the Professional Version of Pamela for Skype - a $5 value! Visit www.pamela-systems.com/store/dummies.php for more information and order Pamela for Skype - Professional Version today! Use promo code PKTH-VA3U-4V when ordering.

**No expiration date.

Pamela for Skype requires the following minimum system requirements:
- PC running Windows 2000, XP or Vista. 32-bit and 64-bit supported. Will not run on Win9x or OSX.
- Skype for Windows, Release 3.0 or higher

Pamela ™

www.pamela-systems.com

Lightning Source UK Ltd.
Milton Keynes UK
UKOW02f1324181113

221311UK00003B/87/P